Exercises and Solutions Manual

Gérard Letac

Exercises and Solutions Manual for

Integration and Probability

by Paul Malliavin

Translated by Leslie Kay

Springer-Verlag
New York Berlin Heidelberg London Paris
Tokyo Hong Kong Barcelona Budapest

Gérard Letac
Laboratoire de Statistique
Université Paul Sabatier
118 Route de Narbonne
F-31062 Toulouse
France

Translator:

Leslie Kay
Department of Mathematics
Virginia Polytechnic Institute
and State University
Blacksburg, VA 24061
USA

Mathematics Subject Classification (1991): 28-01, 43A25, 60H07

Printed on acid-free paper.

Production managed by Frank Ganz; manufacturing supervised by Jeffrey Taub.
Camera-ready copy prepared by the translator.
Printed and bound by Braun-Brumfield, Ann Arbor, MI.
Printed in the United States of America.

9 8 7 6 5 4 3 2 1

ISBN 0-387-94421-4 Springer-Verlag New York Berlin Heidelberg

Contents

I

Measurable Spaces and Integrable Functions

Problem I-1. *If $\mathcal{G}$ is a family of subsets of a set X, we denote by $a(\mathcal{G})$ the Boolean algebra generated by $\mathcal{G}$ and by $\sigma(\mathcal{G})$ the σ-algebra generated by $\mathcal{G}$. A* partition *of X is a family $P = \{P_j\}_{j \in J}$ of nonempty subsets of X such that $P_i \cap P_j = \emptyset$ if $i \neq j$ and $\cup_{j \in J} = X$.*
(1) Let $P = \{P_j\}_{j \in J}$ be a partition of X. Characterize
a) $a(P)$ if J is finite,
b) $a(P)$ if J is infinite,
c) $\sigma(P)$ if J is finite or countable, and
d) $\sigma(P)$ if J is uncountably infinite.
(2) Show that the family $\mathcal{A}$ of subsets of X is a Boolean algebra generated by a finite number of elements if and only if there exists a partition $P = \{P_j\}_{j \in J}$, with J finite, such that $\mathcal{A} = a(P)$.
(3) Let $\mathcal{A}$ be a σ-algebra on a countable set X. Show that there exists a partition P of X such that $\mathcal{A} = \sigma(P)$.
(4) Show that a σ-algebra never has a countable number of elements.

SOLUTION. (1) If $T \subset J$, let $A(T) = \cup_{j \in T} P_j$, with the convention that $A(\emptyset) = \emptyset$.

a) $T \mapsto A(T)$ is a bijection between the set $\mathcal{P}(J)$ of all subsets of J and the set $a(P)$.

- $\mathcal{A} = \{A(T) : T \in \mathcal{P}(J)\}$ is a Boolean algebra which contains P; hence it contains $a(P)$.

- Conversely, it is trivial that $a(P) \supset \mathcal{A}$.

b) Let $\mathcal{F}(J)$ be the set of subsets of J which are either finite or cofinite (complements of finite sets). Then $\mathcal{F}(J)$ is a Boolean algebra. Moreover, $T \mapsto A(T)$ is a bijection between $\mathcal{F}(J)$ and $a(P)$.

- $\mathcal{A} = \{A(T) : T \in \mathcal{F}(J)\}$ is a Boolean algebra which contains P and hence $a(P)$.

- Conversely, $a(P) \supset \mathcal{A}$ trivially.

c) $T \mapsto A(T)$ is a bijection between $\mathcal{P}(J)$ and $\sigma(P)$.

- $\mathcal{A} = \{A(T) : T \in \mathcal{P}(J)\}$ is a σ-algebra which contains P and hence $\sigma(P)$.

- Conversely, $\sigma(P) \supset \mathcal{A}$ trivially.

d) Let $\mathcal{D}(J)$ be the set of subsets of J which are either finite, countable, or cocountable (complements of finite or countable sets). Then $\mathcal{D}(J)$ is a σ-algebra. Moreover, $T \mapsto A(T)$ is a bijection between $\mathcal{D}(J)$ and $\sigma(P)$.

- $\mathcal{A} = \{A(T) : T \in \mathcal{D}(J)\}$ is a σ-algebra which contains P and hence $\sigma(P)$.

- $\sigma(P) \supset \mathcal{A}$ trivially.

(2) Sufficiency is trivial. If $\mathcal{A}$ is generated by $\mathcal{G} = \{G_1, G_2, \ldots, G_n\}$, set $G_j^1 = G_j$ and $G_j^{-1} = G_j^c$. For every $\epsilon = (\epsilon_1, \epsilon_2, \ldots, \epsilon_n)$ with $\epsilon_j = \pm 1$, let $G_\epsilon = \cap_{j=1}^n G_j^{\epsilon_j}$. Let $E = \{\epsilon : G_\epsilon \neq \emptyset\}$. Then $P = \{G_\epsilon\}_{\epsilon \in E}$ is a partition of X, and it is easy to see that $a(\mathcal{G}) \subset a(P)$ and $a(P) \subset a(\mathcal{G})$.
(3) For every x in X, let $A(x) = \cap\{A : A \in \mathcal{A} \text{ and } x \in A\}$ and let an equivalence relation $\sim$ be defined on X by $x \sim y$ if $A(x) = A(y)$. Let $P = \{P_j\}_{j \in J}$ be the partition of X into equivalence classes. (This is just the image of the mapping $x \mapsto A(x)$ from X to $\mathcal{P}(X)$.) We show that $A(x) \in \mathcal{A}$ for every x.

$$(A(x))^c = \{A^c : A \in \mathcal{A} \text{ and } x \in A\} = \{B : B \in \mathcal{A} \text{ and } x \notin B\}.$$

Hence, for every $y \in (A(x))^c$, there exists $B(y)$ such that $y \in B(y)$, $B(y) \in \mathcal{A}$, and $x \notin B(y)$. Thus

$$(A(x))^c = \cup\{B(y) : y \in (A(x))^c\}.$$

This union is countable since X is. Hence $A^c(x) \in \mathcal{A}$ and therefore $\sigma(P) \subset \mathcal{A}$.

Conversely, $A \in \sigma(P)$ for every A in $\mathcal{A}$. For, if not, there would exist $j \in J$ and $x, y \in X$ such that $x \in P_j \cap A$ and $y \in P_j \cap A^c$. Since $x \sim y$ and $y \notin A(x)$, this gives a contradiction. Hence $\mathcal{A} \subset \sigma(P)$.

(4) Let $\mathcal{A}$ be a σ-algebra on X with a countable number of elements. Let $A(x)$, $\sim$, and $P = \{P_j\}_{j\in J}$ be defined as in (3). Then $A(x) \in \mathcal{A}$. But either J is finite and (1) implies that $|A| = 2^{|J|}$, or J is infinite and (1c) and (1d) imply that $\sigma(P)$ is uncountable.

Problem I-2. *Let $\mathcal{G}$ be a family of subsets of a set X such that $X \in \mathcal{G}$ and $\mathcal{G}$ is closed under finite intersections. An* r-family *is a family $\mathcal{R}$ of subsets of X which is closed under finite intersections of pairwise disjoint sets and such that if B_1 and $B_2 \in \mathcal{R}$ with $B_1 \subset B_2$, then $B_2 \setminus B_1 \in \mathcal{R}$. Let $r(\mathcal{G})$ be the smallest r-family containing $\mathcal{G}$. Show that $r(\mathcal{G})$ equals the Boolean algebra $a(\mathcal{G})$ generated by $\mathcal{G}$.*

METHOD. Consider the families

$$\mathcal{R}_1 = \{B : B \in r(\mathcal{G}) \text{ and } A \cap B \in r(\mathcal{G}) \ \forall A \in \mathcal{G}\}$$

$$\mathcal{R}_2 = \{B : B \in r(\mathcal{G}) \text{ and } A \cap B \in r(\mathcal{G}) \ \forall A \in r(\mathcal{G})\}$$

and show that they are r-families.

SOLUTION. It is clear that a Boolean algebra is an r-family and hence that $a(\mathcal{G}) \supset r(\mathcal{G})$. Conversely, it must be shown that $r(\mathcal{G})$ is a Boolean algebra, which will imply that $r(\mathcal{G}) \supset a(\mathcal{G})$. Since $X \in \mathcal{G} \subset r(\mathcal{G})$, clearly $r(\mathcal{G})$ is closed under complements. It remains to show that $r(\mathcal{G})$ is closed under finite intersections.

$\mathcal{R}_1$ and $\mathcal{R}_2$ are obviously closed under disjoint unions. If $B_1 \in \mathcal{R}_i$ and $B_2 \in \mathcal{R}_i$ with $i = 1$ or 2 and $B_1 \subset B_2$, then, for every A,

$$A \cap (B_2 \setminus B_1) = (A \cap B_2) \setminus (A \cap B_1),$$

which implies that $\mathcal{R}_1$ and $\mathcal{R}_2$ are r-families.

Finally, since $\mathcal{G}$ is closed under intersections, $\mathcal{G} \subset \mathcal{R}_1 \subset r(\mathcal{G})$ and thus $\mathcal{R}_1 = r(\mathcal{G})$. It follows that $\mathcal{G} \subset \mathcal{R}_2 \subset r(\mathcal{G})$, which implies that $r(\mathcal{G})$ is closed under intersections.

Problem I-3. *Let $\mathcal{G}_1$ and $\mathcal{G}_2$ be two nonempty families of subsets of a set X which are closed under finite intersections. Let $\mathcal{A}_1$, $\mathcal{A}_2$, and $\mathcal{A}$ denote the σ-algebras generated by $\mathcal{G}_1$, $\mathcal{G}_2$, and $\mathcal{G}_1 \cup \mathcal{G}_2$, respectively. Let P be a measure of total mass 1 on $(X, \mathcal{A})$. Show that if*

$$P(A_1 \cap A_2) = P(A_1)P(A_2) \quad \textit{for all } A_1 \in \mathcal{G}_1 \textit{ and } A_2 \in \mathcal{G}_2,$$

then the same equality holds for all $A_1 \in \mathcal{A}_1$ and $A_2 \in \mathcal{A}_2$.

METHOD. Consider the families

$$\mathcal{M}_1 = \{A : A \in \mathcal{A} \text{ and } P(A \cap A_2) = P(A)P(A_2) \ \forall A_2 \in \mathcal{G}_2\}$$

$$\mathcal{M}_2 = \{A : A \in \mathcal{A} \text{ and } P(A_1 \cap A) = P(A_1)P(A) \ \ \forall A_1 \in \mathcal{A}_1\}$$

and apply the theorem on monotone classes, using Problem I-2.

SOLUTION. It is clear that $\mathcal{M}_1$ and $\mathcal{M}_2$ are monotone classes. By hypothesis, $\mathcal{G}_1 \subset \mathcal{M}_1$. Evidently $X \in \mathcal{M}_1$. Applying the result of Problem 2 to the family $\mathcal{F}_1 = \mathcal{G}_1 \cup \{X\}$, which is closed under finite intersections, we see that $\mathcal{M}_1$ is an r-family; hence

$$\mathcal{M}_1 \supset r(\mathcal{F}_1) = a(\mathcal{F}_1) = a(\mathcal{G}_1).$$

By the theorem on monotone classes, we can write

$$\mathcal{A} \supset \mathcal{M}_1 \supset \mathcal{A}_1, \quad \text{and hence} \quad \mathcal{A} \supset \mathcal{M}_2 \supset \mathcal{G}_2.$$

Similarly, it is clear that $X \in \mathcal{M}_2$. We now apply the result of Problem I-2 to the family $\mathcal{F}_2 = \mathcal{G}_2 \cup \{X\}$, which is closed under finite intersections. Clearly $\mathcal{M}_2$ is an r-family, and hence

$$\mathcal{M}_2 \supset r(\mathcal{F}_2) = a(\mathcal{F}_2) = a(\mathcal{G}_2).$$

By the theorem on monotone classes,

$$\mathcal{A} \supset \mathcal{M}_2 \supset \mathcal{A}_2,$$

which completes the proof.

REMARKS.
1. This result is especially useful in probability theory. Thus, if $X = \mathbf{R}^2$, $A_1(x) = \{(x_1, x_2) : x_1 < x\}$, and $A_2(y) = \{(x_1, x_2) : x_2 < y\}$, then $\mathcal{G}_1 = \{A_1(x) : x \in \mathbf{R}\}$ and $\mathcal{G}_2 = \{A_2(y) : y \in \mathbf{R}\}$ are closed under finite intersections and $\mathcal{A}$ is the set of Borel subsets of $\mathbf{R}^2$. If P is a probability measure on $(\mathbf{R}^2, \mathcal{A})$, it is the distribution of a pair (X_1, X_2) of real random variables. By Problem II-3, (X_1, X_2) is a pair of independent random variables if and only if

$$P[X_1 < x;\ X_2 < y] = P[X_1 < x] \cdot P[X_2 < y]$$

for all $(x, y) \in \mathbf{R}^2$.
2. The result can be extended from two factors to n factors by constructing monotone classes $\mathcal{M}_k$ for $k = 1, 2, \ldots, n$ and using induction on k. One proves first that $\mathcal{M}_k \supset \mathcal{G}_k$, then that $\mathcal{M}_k \supset \mathcal{A}_k$.

Problem I-4. *Let $x = \{x_n\}_{n=0}^{\infty}$ and let*

$$\ell^\infty = \left\{ x : x_n \in \mathbf{R} \ \ \forall n \in \mathbf{N} \ \ \textit{and} \ \ \|x\|_\infty = \sup_n |x_n| < \infty \right\}.$$

Define $T : \ell^\infty \to \ell^\infty$ by $(Tx)_0 = x_0$ and $(Tx)_n = x_n - x_{n-1}$ if $n > 0$.

(1) If $e = (1,1,\ldots,1,\ldots)$, show that the equation $Tx = e$ has no solution x in ℓ^∞.
(2) Let $F = T\ell^\infty$ be the image of T. Assume without proof that there exists a continuous linear functional f on ℓ^∞ such that $f(x) = 0$ for every x in F, $f(e) = 1$, and $\sup\{|f(x)| : \|x\|_\infty \leq 1\} < +\infty$ (Hahn-Banach Theorem). Show that if $x = \{x_n\}_{n=0}^\infty$ is such that $x_n \geq 0$ for every n, then $f(x) \geq 0$.
(3) Let $S : \ell^\infty \to \ell^\infty$ be defined by $(Sx)_n = x_{n+1}$ if $n \geq 0$. Show that $f(x) = f(Sx)$ for every x in ℓ^∞.
(4) Show that $\liminf_{n\to+\infty} x_n \geq 0$ implies that $f(x) \geq 0$. Conclude that $\liminf_{n\to+\infty} x_n \leq f(x) \leq \limsup_{n\to+\infty} x_n$ for every $x \in \ell^\infty$.
(5) Let $A \subset \mathbf{N}$ and let $\mathbf{1}_A \in \ell^\infty$ be defined by $\mathbf{1}_A(n) = 0$ if $n \neq A$ and $\mathbf{1}_A(n) = 1$ if $n \in A$. If $P(A) = f(\mathbf{1}_A)$, show that $P(A \cup B) = P(A) + P(B)$ if $A \cap B = \emptyset$ and that P does not satisfy the countable additivity axiom.

SOLUTION. (1) $Tx = e$ implies that $x_0 = 1$ and, by induction on n, that $x_n = n + 1$. Hence $e \neq T\ell^\infty$.
(2) Assume that $f(x) < 0$. Then $\|x\|_\infty > 0$; if $y = e - \frac{x}{\|x\|_\infty}$, with $y_n = 1 - \frac{x_n}{\|x_n\|_\infty}$, then $0 \leq y_n \leq 1$. Hence $|f(y)| \leq 1$, but $f(y) = 1 - \frac{f(x)}{\|x\|} > 1$, a contradiction.
(3) In order to see that $f(x) = f(Sx)$, it suffices to show that $x - Sx \in F$, and hence that for every x the equation $x - Sx = Ty$ has a solution y. Indeed, we find that $y_n = x_0 - x_{n+1}$ for every n, and this defines an element y of ℓ^∞.
(4) If $\liminf_{n\to+\infty} x_n \geq 0$, then for every $\epsilon > 0$ there exists an integer $N(\epsilon)$ such that $x_k \geq -\epsilon$ if $k \geq N(\epsilon)$. Thus $S^{N(\epsilon)}x + \epsilon e)_n \geq 0$ for every n; by (2), $f(S^{N(\epsilon)}x + \epsilon e) \geq 0$ and hence $f(S^{N(\epsilon)}x) \geq -\epsilon$. By (3), we have $f(x) = f(S^{N(\epsilon)}x)$, whence $f(x) \geq -\epsilon$ for every $\epsilon > 0$, or $f(x) \geq 0$.

Moreover, if x is an arbitrary element of ℓ^∞, let $m = \liminf_{n\to+\infty} x_n$ and let $M = \limsup_{n\to+\infty} x_n$. Then $f(x - me)$ and $f(Me - x)$ are nonnegative by the first paragraph of (4), and hence $m \leq f(x) \leq M$.
(5) That $P(A \cup B) = P(A) + P(B)$ if $A \cap B = \emptyset$ follows immediately from the linearity of f. To see that P does not satisfy the countable additivity axiom, consider $P(\{k\}) = f(\mathbf{1}_{\{k\}})$. Since $\lim_{n\to+\infty} \mathbf{1}_{\{k\}}(n) = 0$, it follows from (4) that $P(\{k\|) = 0$, and hence that

$$1 = f(e) = P(\mathbf{N}) = P(\cup_{k\in\mathbf{N}}\{k\}) \neq \sum_{k\in\mathbf{N}} P(\{k\}) = 0.$$

REMARKS. The linear functional f above is called a *Banach limit*; it cannot be written down explicitly since it is constructed by means of the Hahn-Banach theorem and the axiom of choice. Similarly, it is impossible to give an explicit example of an additive but not σ-additive measure on a σ-algebra.

Problem I-5. *Let X be an uncountable set and let $\mathcal{A}$ be the σ-algebra generated by the family of* 1*-element subsets of X. (See Problem 1, question (1d).) Let $P : \mathcal{A} \to [0, 1]$ be defined by*

$$\begin{array}{lll} P(A) = 0 & \textit{if} & A \textit{ is finite or countable} \\ P(A) = 1 & \textit{if} & A \textit{ is cocountable.} \end{array}$$

(A is cocountable if A^c is finite or countable.) Show that P is a probability measure on $(X, \mathcal{A})$.

SOLUTION. Let $\{A_n\}_{n=0}^{\infty}$ be a sequence of pairwise disjoint elements of $\mathcal{A}$. Then

- either, for every n, A_n is finite or countable, in which case so is $\cup_{n=0}^{\infty} A_n$ and hence

$$P(\cup_{n=0}^{\infty} A_n) = 0 = \sum_{n=0}^{\infty} 0 = \sum_{n=0}^{\infty} P(A_n);$$

- or there exists $n_0 \in \mathbf{N}$ such that A_{n_0} is cocountable. Since A_n is disjoint from A_{n_0} if $n \neq n_0$, $\cup_{n \neq 0} A_n$ is finite or countable. Hence

$$P(\cup_{n=0}^{\infty} A_n) = 1 = P(A_{n_0}) + P(\cup_{n \neq n_0} A_n) = \sum_{n=0}^{\infty} P(A_n).$$

Problem I-6. *Let $(X, \mathcal{A}, \mu)$ be a measure space and let f be a nonnegative measurable function on X. For every $t \geq 0$, set*

$$F(t) = \mu\{x : f(x) > t\} \quad \textit{and} \quad G(t) = \mu\{x : f(x) \geq t\}.$$

(1) Assume that $f(X) \subseteq \mathbf{N}$ and that f is integrable. Prove that

$$\int_X f(x) d\mu(x) = \sum_{n=0}^{\infty} F(n) = \sum_{n=1}^{\infty} G(n).$$

METHOD. Set $\mu_n = \mu\{x : f(x) = n\}$ and show that $\int_X f(x)\mu(dx) = \sum_{n=0}^{\infty} n\mu_n$.

(2) Assume that f^{α} is integrable for $\alpha > 0$. Prove that

$$\int_X f^{\alpha}(x) d\mu(x) = \alpha \int_0^{+\infty} t^{\alpha-1} F(t) dt = \alpha \int_0^{+\infty} t^{\alpha-1} G(t) dt.$$

METHOD. Show that (2) holds for $\alpha = 1$ by considering the functions $f_n(x) = \frac{[2^n f(x)]}{2^n}$, where $[a]$ means "the greatest integer $\leq a$", and using the monotone convergence theorem. The general case can then be reduced to the case $\alpha = 1$.

SOLUTION. (1) If $f_k(x) = x$ for $x \leq k$ and $f_k(x) = 0$ for $x > k$, f_k is a simple function (one which assumes only finitely many values) and

$$\int_X f_k(x)d\mu(x) = 0 \times \mu(\{x : x > k\}) + \sum_{n=0}^{k} n\mu_n.$$

Since f_k approaches f as $k \to +\infty$, f is integrable, and $0 \leq f_k \leq f$, the dominated convergence theorem (I-7.6) implies that

$$\int_X f_k(x)\mu(dx) = \lim_{k\to+\infty} \int_X f_k(x)\mu(dx) = \lim_{k\to+\infty} \sum_{n=0}^{k} n\mu_n = \sum_{n=0}^{\infty} n\mu_n.$$

This implies that $F(n) < +\infty$. Observe next that

$$\sum_{n=0}^{N} n\mu_n = \sum_{n=0}^{N-1} (F(n) - F(N)) \leq \sum_{n=0}^{N-1} F(n) = \sum_{n=0}^{\infty} \min(n, N)\mu_n \leq \sum_{n=0}^{\infty} n\mu_n.$$

Hence

$$\sum_{n=0}^{\infty} n\mu_n = \sum_{n=0}^{\infty} F(n).$$

Since $F(n) = \int_n^{n+1} F(t)dt$, the first equality has been proved. The second is clear, since $G(n+1) = F(n)$ for every $n \geq 0$.

(2) We begin by proving the second equality. Let $A(t) = \{x : f(x) = t\}$. The $A(t)$ are disjoint and, *since μ is σ-finite,* $D = \{t : \mu(A(t)) > 0\}$ cannot be uncountable. The functions $t^{\alpha-1}F(t)$ and $t^{\alpha-1}G(t)$ coincide on $[0, +\infty) \setminus D$ and their integrals are equal.

Assume now that $\alpha = 1$.

Since $2^n f_n$ has integer values, it follows from (1) that

$$\int_X f_n(x)d\mu(x) = \frac{1}{2^n} \sum_{k=1}^{\infty} G\left(\frac{k}{2^n}\right).$$

Since G is decreasing, we have the following bounds

$$\sum_{k=1}^{\infty} \frac{1}{2^n} G\left(\frac{k}{2^n}\right) \leq \int_0^{+\infty} G(x)dx \leq \sum_{k=0}^{\infty} \frac{1}{2^n} G\left(\frac{k}{2^n}\right).$$

Moreover, $f_n(x)$ increases to $f(x)$, and hence $\int_X f_n(x)d\mu(x) \int_X f(x)d\mu(x)$ as $n \to +\infty$ by the monotone convergence theorem. This implies the result for $\alpha = 1$.

Finally, if $\alpha > 0$, we can write

$$\int_X f^\alpha(x)d\mu(x) = \int_0^\infty \mu\{x : f^\alpha(x) > u\}du = \alpha \int_0^\infty t^{\alpha-1}F(t)dt,$$

where the last equality is obtained by the change of variable $u = t^\alpha$ and the first by applying the result for $\alpha = 1$ to the function f^α rather than f.

Problem I-7. *For $0 < r < 1$, we write the Poisson kernel as*

$$P_r(\theta) = 1 + 2\sum_{n=1}^\infty r^n \cos n\theta = \frac{1-r^2}{1-2r\cos\theta + r^2}.$$

(1) Show that $r^2 + \cos\theta(1-2r) \geq 0$ if $0 \leq \theta \leq \pi$ and $\frac{1}{2} \leq r \leq 1$. Deduce that $\theta^2 P_r(\theta) \leq \frac{(1-r^2)\theta^2}{1-\cos\theta}$ and evaluate $\lim_{r\to 1}\int_0^\pi \theta^2 P_r(\theta)d\theta$.
(2) Show that $\int_0^\pi \theta^2 P_r(\theta)d\theta = \frac{\pi^3}{3} + 4\pi\sum_{n=1}^\infty \frac{(-r)^n}{n^2}$ and use this to derive another expression for $\lim_{r\to 1}\int_0^\pi \theta^2 P_r(\theta)d\theta$.
(3) Use (1) and (2) to compute the sums of the series $\sum_{n=1}^\infty \frac{(-1)^n}{n^2}$, $\sum_{n=1}^\infty \frac{1}{(2n-1)^2}$, and $\sum_{n=1}^\infty \frac{1}{n^2}$.
(4) Express $\int_0^1 (\log(1-x))^2 \frac{dx}{x^2}$ as the sum of a double series and show that $\int_0^1 (\log(1-x^2))^2 \frac{dx}{x^2} = 2\sum_{n=1}^\infty \frac{1}{n^2}$.

SOLUTION.
(1) $0 \leq (1-r)^2 = r^2 + 1 - 2r \leq r^2 + \cos\theta(1-2r)$. The inequality $\theta^2 P_r(\theta) \leq \frac{(1-r)^2\theta^2}{1-\cos\theta}$ is immediate. Since $P_r(\theta) \geq 0$ and $\frac{\theta^2}{1-\cos\theta}$ is integrable, it follows from the dominated convergence theorem that

$$\int_0^\pi \frac{\theta^2 d\theta}{1-2r\cos\theta + r^2} \to \frac{1}{2}\int_0^\pi \frac{\theta^2 d\theta}{1-\cos\theta} \quad \text{as} \quad r \to 1.$$

Hence $\lim_{r\to 1}\int_0^\pi \theta^2 P_r(\theta) = 0$.
(2)

$$\int_0^\pi \theta^2 P_r(\theta)d\theta = \int_0^\pi \theta^2 d\theta + 2\sum_{n=1}^\infty r^n \int_0^\pi \theta^2 \cos\ n\theta\ d\theta$$
$$= \frac{\pi^3}{3} + 2\sum_{n=1}^\infty r^n \left[\frac{\theta^2}{n}\sin\ n\theta + \frac{2}{n^2}\theta\cos\ n\theta - \frac{2}{n^3}\sin\ n\theta\right]_0^\pi.$$

The first equality is justified by the fact that if $s_N(\theta) = \theta^2 + \sum_{n=1}^N r^n\theta^2 \cos\ n\theta$, then $|s_N(\theta)| \leq \theta^2(1 + 2r\frac{1-r^N}{1-r})$ for fixed r in $(0,1)$. Hence $s_N(\theta) \leq \theta^2\frac{1+r}{1-r}$. Thus the dominated convergence theorem can be applied to the sequence $\{s_N\}_{N=1}^\infty$. Furthermore, since $|(-r)^n| \leq 1$, applying dominated convergence again shows that

$$\lim_{r\to 1}\int_0^\pi \theta^2 P_r(\theta)d\theta = \frac{\pi^3}{3} + 4\pi\sum_{n=1}^\infty \frac{(-1)^n}{n^2}.$$

(3) From (1) and (2) it follows that $\frac{\pi^3}{3} + 4\pi \sum_{n=1}^{\infty} \frac{(-1)^n}{n^2} = 0$ and hence that $\sum_{n=1}^{\infty} \frac{(-1)^n}{n^2} = -\frac{\pi^2}{12}$. Let $I = \sum_{k=1}^{\infty} \frac{1}{(2k-1)^2}$ and let $P = \sum_{k=1}^{\infty} \frac{1}{(2k)^2}$. It is easy to see that $I + P = 4P$ and, by the result above, $P - I = -\frac{\pi^2}{12}$. Hence $I = \frac{\pi^2}{8}$, $P = \frac{\pi^2}{24}$, and $\sum_{n=1}^{\infty} \frac{1}{n^2} = \frac{\pi^2}{6}$.

(4) $\frac{1}{x} \log(1-x) = \sum_{n=0}^{\infty} \frac{x^n}{n+1}$ for $0 < x < 1$. Since this is a convergent series with positive terms, the double series $\sum_{n=0}^{\infty} \sum_{m=0}^{\infty} \frac{x^{n+m}}{(n+1)(m+1)}$ is convergent with positive terms and has sum $(\log(1-x))^2 \frac{1}{x^2}$. Thus, by monotone convergence,

$$\int_0^1 (\log(1-x))^2 \frac{dx}{x^2} = \sum_{n=0}^{\infty} \sum_{m=0}^{\infty} \frac{1}{(n+m+1)(n+1)(m+1)}.$$

If $n \neq 0$, we can use partial fractions to re-write the summands as

$$\frac{1}{(n+m+1)(n+1)(m+1)} = \frac{1}{(n+1)n} \left[\frac{1}{m+1} - \frac{1}{n+m+1} \right],$$

and hence

$$\begin{aligned}
&\sum_{n=1}^{\infty} \sum_{m=0}^{\infty} \frac{1}{(n+m+1)(n+1)(m+1)} \\
&\quad = \sum_{n=1}^{\infty} \frac{1}{(n+1)n} \sum_{m=0}^{\infty} \left(\frac{1}{m+1} - \frac{1}{n+m+1} \right) \\
&\quad = \sum_{n=1}^{\infty} \frac{1}{(n+1)n} \sum_{m=1}^{\infty} \frac{1}{m} \\
&\quad = \sum_{m=1}^{\infty} \frac{1}{m} \sum_{n=m}^{\infty} \frac{1}{(n+1)n} \\
&\quad = \sum_{m=1}^{\infty} \frac{1}{m} \sum_{n=m}^{\infty} (\frac{1}{n} - \frac{1}{n+1} \\
&\quad = \sum_{m=1}^{\infty} \frac{1}{m^2}.
\end{aligned}$$

The $n = 0$ term in the double series also gives $\sum_{m=0}^{\infty} \frac{1}{(m+1)^2}$. Thus

$$\int_0^1 (\log(1-x))^2 \frac{dx}{x^2} = 2 \sum_{n=1}^{\infty} \frac{1}{n^2} = \frac{\pi^2}{3}.$$

Problem I-8. *Evaluate* $\sum_{n=1}^{\infty} \frac{(-1)^{n+1}}{n}$ *by using the integral* $\int_0^1 \frac{dx}{1+x}$ *and the monotone convergence theorem.*

SOLUTION.

$$\frac{1}{1+x} = \sum_{n=0}^{\infty}(-x)^n = \sum_{n=0}^{\infty} x^{2n}(1-x) \quad \text{for} \quad 0 < x < 1.$$

Since the second series has positive terms, the monotone convergence theorem can be applied to give

$$\log 2 = \int_0^1 \frac{dx}{1+x} = \sum_{n=0}^{\infty}(\frac{1}{2n+1} - \frac{1}{2n+2}) = \sum_{n=1}^{\infty} \frac{(-1)^{n+1}}{n}.$$

REMARK. Even an example as simple as this shows how much easier it is to work with the Lebesgue integral than the Riemann integral. Justifying the preceding calculation is much more delicate when $\int_0^1 \frac{dx}{1+x}$ is viewed as a Riemann integral.

Problem I-9. *Let $(X, \mathcal{A}, \mu)$ be a measure space and let $x \mapsto f(x) = (f_1(x), f_2(x), \dots, f_n(x))$ be a measurable mapping from X to $\mathbf{R}^n$. Suppose that $\mathbf{R}^n$ is equipped with a norm $\| \ \|$ such that $x \mapsto \|f(x)\|$ is integrable.*
(1) Show that f_j is integrable for every $j = 1, 2, \dots, n$.
(2) Defining $\int_X f(x) d\mu(x)$ in $\mathbf{R}^n$ by

$$\left(\int_X f_1(x)\mu(dx), \dots, \int_X f_1(x)\mu(dx)\right),$$

show that $\| \int_X f(x)\mu(dx)\| \le \int_X \|f(x)\|\mu(dx)$.

METHOD. On the dual space $(\mathbf{R}^n)^*$ consisting of linear functionals $a : v \mapsto \langle a, v\rangle$ on $\mathbf{R}^n$, introduce the dual norm $\|a\|^* = \sup_{v \neq 0} \frac{|\langle a,v\rangle|}{\|v\|}$ and use the fact that $\|v\| = \sup_{a \neq 0} \frac{|\langle a,v\rangle|}{\|a\|^*}$.

SOLUTION.
(1) If $v = (v_1, v_2, \dots, v_n)$, the mapping $v \mapsto v_j$ from $\mathbf{R}^n$ to $\mathbf{R}$ is linear and hence continuous, since it is a map between finite-dimensional spaces. Hence there exists a number $k > 0$ such that $|v_j| \le k\|v\|$ for every v in $\mathbf{R}^n$. Thus

$$\int_X |f_j(x)|\mu(dx) \le k \int_X \|f(x)\|\mu(dx).$$

(2) If $a \in (\mathbf{R}^n)^*$, we can write $|\langle a, f(x)\rangle| \le \|f(x)\| \ \|a\|^*$, and hence

$$\langle a, \int_X f(x)\mu(dx)\rangle = \int_X \langle a, f(x)\rangle \mu(dx) \le \|a\|^* \int_X \|f(x)\|\mu(dx).$$

Taking the sup over a and using the fact mentioned above, we find that

$$\| \int_X f(x)\mu(dx) \| \leq \int_X \|f(x)\| \mu(dx).$$

REMARKS.

1. *The shortest path between two points is a straight line.* Consider $\mathbf{R}^n$ with the Euclidean norm $\|v\| = \left[v_1^2 + v_2^2 + \cdots v_n^2\right]^{\frac{1}{2}}$. Let $X = [0,1]$ with Lebesgue measure. (See Chapter II.) Let F be a function from $[0,1]$ to $\mathbf{R}^n$ such that the derivative $f = F'$ exists everywhere and is continuous. Then $\int_0^1 \|f(x)\|dx$ can be interpreted as the Euclidean length of the curve described by F, and $\| \int_0^1 f(x)dx\| = \|F(1) - F(0)\|$ is the length of the line segment with endpoints $F(0)$ and $F(1)$.

2. *Case of equality.* It can be shown that, when the unit ball B is strictly convex (that is, when $\|v_1\| = \|v_2\| = \|\lambda v_1 + (1-\lambda)v_2\| = 1$ for $\lambda \in [0,1]$ holds only for $\lambda = 0$ or 1), the inequality is strict unless there exist $v \in \mathbf{R}^n$ and a function $g(x) \geq 0$ such that $f(x) = g(x)v$ μ-almost everywhere. The application to the Euclidean length of a curve is immediate.

Problem I-10. *Let $X, X_1, \ldots, X_n, \ldots$ be measurable functions from a space $(E, \mathcal{E}, \mu)$ to an open set Ω of Euclidean space $\mathbf{R}^d$ such that*

$$\forall \epsilon > 0 \quad \mu(\{\|X_n - X\| \geq \epsilon\}) \to 0 \quad \text{as } n \to \infty.$$

(1) Show that $\forall \epsilon > 0$ there exists a compact set $K \subset \Omega$ such that $\mu(\{X \notin K\}) \leq \epsilon$ and, for every n, $\mu(\{X_n \notin K\}) \leq \epsilon$.
(2) If $f : \Omega \to \mathbf{R}^m$ is continuous, then $\forall \epsilon > 0$

$$\mu(\{\|f(X_n) - f(X)\| \geq \epsilon\}) \to 0 \quad \text{as} \quad n \to \infty.$$

SOLUTION.

(1) For any fixed $\epsilon > 0$, there exists a compact subset C_1 of Ω such that $\mu(\{X \notin C_1\}) \leq \frac{\epsilon}{2}$. Let $a = \inf\{\|x - y\| : x \in \Omega \text{ and } y \in C_1\}$ and let $0 < b < a$ be fixed. Then there exists an integer N such that $\mu(\{\|X_n - X\| \geq b\}) \leq \frac{\epsilon}{2}$ if $n \geq N$. Let C_2 be the compact subset of Ω defined by

$$C_2 = \{x \in \Omega : \inf\{\|x - y\| : y \in C_1\} \leq b\}.$$

Then, if $n \geq N$,

$$\begin{aligned} \mu(\{X_n \notin C_2\}) &= \mu(\{X_n \notin C_2\} \cap \{\|X_n - X\| \leq b\}) \\ &\quad + \mu(\{X_n \notin C_2\} \cap \{\|X_n - X\| \geq b\}) \\ &\leq \mu(\{X \notin C_1\}) + \mu(\{\|X_n - X\| \geq b\}) \leq \epsilon. \end{aligned}$$

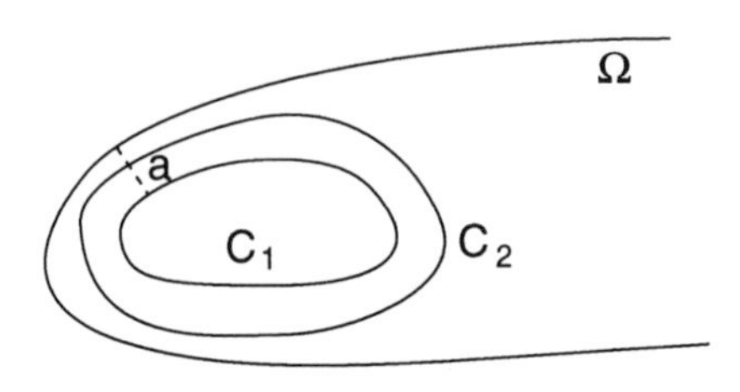

Moreover, if $n < N$, there exists a compact subset K_n of Ω such that $\mu(\{X_n \notin K_n\}) \leq \epsilon$. Set $K = C_2 \cup K_0 \cup \ldots \cup K_{N-1}$; the compact set K has the desired property.

(2) We may assume without loss of generality that $m = 1$ since, if $f_j(x)$ is the jth component of f,

$$\mu(\{\|f(X_n) - f(X)\| \geq \epsilon\}) \leq \sum_{j=1}^{m} \mu(\{\|f_j(X_n) - f(X)\| \geq \epsilon\}).$$

Now let $\delta > 0$. By (1), there exists a compact set K in Ω such that $\mu(\{X_n \notin K\}) \leq \delta$ for every n and $\mu(\{X \notin K\}) \leq \delta$. The function f restricted to K is uniformly continuous; that is, for every $\epsilon > 0$ there exists $\eta > 0$ such that $|f(x) - f(y)| \leq \epsilon$ if $\|x - y\| \leq \eta$ with x and y in K. Hence, for every n,

$$\begin{aligned}
&\mu(\{|f(X_n) - f(X)| \geq \epsilon\}) \\
&\quad \leq \mu(\{X_n \in K,\ X \in K,\ |f(X_n) - f(X)| \geq \epsilon\}) \\
&\qquad + \mu(\{X_n \notin K\}) + \mu(\{X \notin K\}) \\
&\quad \leq \mu(\{\|X_n - X\| \geq \eta\}) + 2\delta,
\end{aligned}$$

and therefore

$$\limsup_{n\to\infty} \mu(\{|f(X_n) - f(X)| \geq \epsilon\}) \leq 2\delta.$$

Since δ is arbitrary, $\lim_{n\to\infty} \mu(\{|f(X_n) - f(X)| \geq \epsilon\}) = 0$.

Problem I-11. *Let $(X, \mathcal{A}, \mu)$ and $(Y, \mathcal{B}, \nu)$ be measure spaces such that $\mu(X)$ and $\nu(Y) > 0$. Let $a : X \to \mathbf{C}$ and $b : Y \to \mathbf{C}$ be functions, respectively $\mathcal{A}$ and $\mathcal{B}$ measurable, such that*

$$a(x) = b(y) \quad \mu \otimes \nu\text{-almost everywhere on } X \times Y.$$

Show that there exists a constant λ such that $a(x) = \lambda$ μ-a.e. and $b(y) = \lambda$ ν-a.e.

Solution. If $y \in Y$, let $A_y = \{x : a(x) \neq b(y)\}$ and let $C = \cup_{y \in Y} A_y$. Clearly $A_y \in \mathcal{A}$ and $C \in \mathcal{A} \otimes \mathcal{B}$ (since $(x, y) \mapsto a(x) - b(y)$ is $\mathcal{A} \otimes \mathcal{B}$ measurable). Then $b(y_1) = b(y_2)$ for any y_1, $y_2 \notin B$. For otherwise we would have $a(x) = b(y_1)$ on $A_{y_1}^c$ and $a(x) = b(y_2)$ on $A_{y_2}^c$, with $A_{y_1}^c$ and $A_{y_2}^c$ disjoint and $A_{y_1} \cup A_{y_2} = X$. But this is impossible since $\mu(A_{y_1}) = \mu(A_{y_2}) = 0$ and

$\mu(X) > 0$. Hence b equals a constant λ on B^c. Since $\nu(B) = 0$, $b(y) = \lambda$ ν-a.e. By symmetry, there exists a constant λ' such that $a(x) = \lambda'$ μ-a.e., and it is clear that $\lambda = \lambda'$.

Problem I-12. *On a measure space $(X, \mathcal{A}, \mu)$, let f and g be complex functions such that $|f|^2$ and $|g|^2$ are μ-integrable. Consider the function*

$$h(x,y) = |f(x)g(y) - f(y)g(x)|^2.$$

(1) Show that $0 \le \int_{X\times X} h(x,y)d\mu(x)\, d\mu(y)$, and use this to prove Schwarz's inequality:

$$\left|\int_X f(x)\overline{g(x)}d\mu(x)\right|^2 \le \int_X |f(x)|^2 d\mu(x) \int_X |g(x)|^2 d\mu(x).$$

METHOD. Consider first the case where $f \ge 0$ and $g \ge 0$.

(2) Show that equality holds in Schwarz's inequality if and only if either $g(x) = 0$ μ-a.e. on X or there exists a constant $\lambda \in \mathbf{C}$ such that $f(x) - \lambda g(x) = 0$ μ-a.e. on X.

METHOD. Problem I-11 can be used.

SOLUTION.
(1) Decompose h as follows:

$$(i) \qquad \begin{aligned} h(x,y) = \; & |f(x)|^2|g(y)|^2 + |g(x)|^2|f(y)|^2 \\ & - f(x)\overline{g}(x)\overline{f}(y)g(y) - \overline{f}(x)g(x)f(y)\overline{g}(y). \end{aligned}$$

Now assume f, $g \ge 0$. In this case, by (i),

$$0 \le h(x,y) \le f^2(x)g^2(y) + g^2(x)f^2(y),$$

and by Fubini the right-hand side is integrable with respect to $\mu\otimes\mu$. Hence the same is true of h. Integrating (i) with respect to $\mu\otimes\mu$ immediately gives Schwarz's inequality for $f \ge 0$ and $g \ge 0$.

Now take f and g complex. Schwarz's inequality, applied to $|f|$ and $|g|$, shows that $\int |f(x)\overline{g}(x)|d\mu(x) < \infty$. It follows from (i) and Fubini that h is a linear combination of four functions which are integrable with respect to $\mu\otimes\mu$. Integrating (i) with respect to $\mu\otimes\mu$ and using Fubini, we obtain

$$(ii) \qquad \begin{aligned} & 2\left[\int_X |f(x)|^2 d\mu(x) \int_X |g(x)|^2 d\mu(x) - \left|\int_X f(x)\overline{g(x)}d\mu(x)\right|^2\right] = \\ & \int_{X\times X} h(x,y)d\mu(x)\, d\mu(y) \ge 0. \end{aligned}$$

(2) It follows from (ii) that equality holds for the pair (f, g) if and only if

$$f(x)g(y) = g(x)f(y) \quad \mu \otimes \mu\text{-a.e. on } X \times X.$$

If $g(x) = 0$ μ-a.e., equality holds. Otherwise $\mu(A) \neq 0$, where $A = \{x : g(x) \neq 0\}$. Let $A_1 = \{x : x \in A^c \text{ and } f(x) \neq 0\}$. If $\mu(A_1) > 0$, then, for $(x, y) \in A_1 \times A$,

$$f(x)g(y) - f(y)g(x) = f(x)g(y) \neq 0.$$

But this is impossible since $(\mu \otimes \mu)(A_1 \times A) = \mu(A_1)\mu(A) > 0$. Hence $f(x) = 0$ on A^c μ-a.e.

We now have $\frac{f(x)}{g(x)} = \frac{f(y)}{g(y)}$ $\mu \otimes \mu$-a.e. on $A \times A$, which implies by Problem I-11 that there exists λ such that $\frac{f(x)}{g(x)} = \lambda$ μ-a.e. on A; hence $f(x) = \lambda g(x)$ μ-a.e. on X.

The converse is clear.

REMARK. This method of proving Schwarz's inequality is interesting because it allows the difference between the right-hand and left-hand sides to be evaluated explicitly as the integral of a nonnegative function.

Problem I-13. *If X and Y are measurable real-valued functions defined on the measure space $(\Omega, \mathcal{A}, \mu)$ such that $\mu(\{Y \leq x < X\}) = 0$ for all real x, show that $\mu(\{Y < X\}) = 0$.*

SOLUTION. If $\mu(\{X - Y > 0\}) > 0$, then there exists a number $\epsilon > 0$ such that $\mu(\{X - Y > \epsilon\}) > 0$ since $\cup_{n=1}^{+\infty}\{X - Y > \frac{1}{n}\} = \{X - Y > 0\}$. Hence $\{X - Y > \epsilon\} \subset \cup_{n=1}^{+\infty}\{y \leq n\epsilon < X\}$ and

$$0 < \mu(\{X - Y > \epsilon\}) \leq \sum_{n=1}^{+\infty} \mu(\{Y \leq n\epsilon < X\}) = 0.$$

This gives a contradiction.

Problem I-14. *Let $(X, \mathcal{A}, \mu)$ be a measure space, where $\mu(X)$ is not necessarily finite, let $(Y, \mathcal{B})$ be a measurable space, and let f be a measurable mapping from X to Y. Suppose that there exists a sequence $\{B_n\}$ in $\mathcal{B}$ such that $\cup_{n=1}^{\infty} B_n = y$ and $\mu(f^{-1}(B_n)) < \infty$ for every n.*
(1) Show that $\nu(B) = \mu(f^{-1}(B))$ defines a measure ν on $(Y, \mathcal{B})$ (called the image of μ under f).
(2) Show that if $g \in L^1(\nu)$, then

$$\int_X g(f(x))\mu(dx) = \int_Y g(y)\nu(dy).$$

SOLUTION. (1) It is trivial to check that ν satisfies axioms I-3.0.1 and I-3.0.2.
(2) We show that the formula holds when g is a simple function: $g = \sum_{i=1}^{n} \alpha_i \mathbf{1}_{B_i}$, with $B_i \in \mathcal{B}$ and $\mu(B_i) < \infty$ for $i = 1, 2, \ldots, n$. In this case,

$$\begin{aligned} \int_Y g(y)\nu(dy) &= \sum_{i=1}^{n} \alpha_i \nu(B_i) = \sum_{i=1}^{n} \alpha_i \mu(f^{-1}(B_i)) \\ &= \sum_{i=1}^{n} \alpha_i \int_X \mathbf{1}_{B_i}(f(x))dx \\ &= \int_X g(f(x))\mu(dx). \end{aligned}$$

The result can be extended to $L^1(\nu)$ by density.

REMARKS. 1. The image measure always exists when μ is bounded; this is used extensively in probability theory, in Chapter IV. It does not always exist if $\mu(X) = +\infty$. For example, if $X = \mathbf{R}^2$ is equipped with Lebesgue measure $\mu = dx\, dy$ and $f : \mathbf{R}^2 \to \mathbf{R} = Y$ is the projection $f(x, y) = x$, the image of μ does not exist.
2. If X and Y are metrizable locally compact spaces that are countable at infinity and μ is a Radon measure on X, a sufficient condition for existence of the image measure is that, for every compact set K in Y, $f^{-1}(K)$ should be relatively compact. See Problems II-11, II-12, II-13, and III-3.

Problem I-15. *(1) Let f be square integrable on $[0, 1]$ and let $F(x) = \int_0^x f(t)dt$. Applying the Schwarz inequality to the product $f \times 1$ on $[0, x]$, show that* $\lim_{x \downarrow 0} x^{-\frac{1}{2}} F(x) = 0$.
(2) Let g be square integrable on $[0, +\infty)$ and let $G(x) = \int_0^x g(t)dt$. Applying the Schwarz inequality to the product $g \times 1$ on $[a, x]$, with a sufficiently large, show that $\lim_{x \to +\infty} x^{-\frac{1}{2}} G(x) = 0$.

SOLUTION. (1) $|F^2(x)| \le \int_0^x |f^2(t)|dt \times \int_0^x dt = x \int_0^x |f^2(t)|dt$. Since the function $x \mapsto \int_0^x |f^2(t)|dt$ is continuous, the result follows.
(2) $|G(x) - G(a)|^2 \le (x - a) \int_0^x |g^2(t)|dt$. Hence

$$\frac{|G(x)|}{\sqrt{x}} \le \frac{|G(a)|}{\sqrt{x}} + \sqrt{1 - \frac{a}{x}} \left[\int_a^\infty |g^2(t)|dt\right]^{\frac{1}{2}}.$$

For a given $\epsilon > 0$, first choose a such that $\int_a^\infty |g^2(t)|dt \le \frac{\epsilon^2}{4}$, then choose $X(\epsilon) > a$ such that $\frac{|G(a)|}{\sqrt{X(\epsilon)}} \le \frac{\epsilon}{2}$. Then $\frac{|G(x)|}{\sqrt{X(\epsilon)}} \le \epsilon$ if $x \ge X(\epsilon)$.

REMARK. It is easy to replace L^2 by L^p, with $p > 1$. If $\frac{1}{p} + \frac{1}{q} = 1$, we find that $x^{-\frac{1}{q}} F(x) \to 0$ as $x \to 0$ and $x^{-\frac{1}{q}} G(x) \to 0$ as $x \to +\infty$.

II

Borel Measures and Radon Measures

Problem II-1. *Let I be an* open *interval in* $\mathbf{R}$, *equipped with the Borel algebra* $\mathcal{B}$. *A function* $F : I \to \mathbf{R}$ *is called increasing if* $x < y$ *implies that* $F(x) \leq F(y)$. *We set* $F(x-0) = \lim_{y \uparrow x} F(y)$, $F(x+0) = \lim_{y \downarrow x} F(y)$, *and* $D_F = \{x : F(x-0) \neq F(x+0)\}$.
(1) If $F : I \to \mathbf{R}$ *is increasing, prove that* D_F *is finite or countable.*
METHOD. If $[a,b] \subset I$, show that $D(n;[a,b]) = \{x \in [a,b] : F(x+0) - F(x-0) \geq \frac{1}{n}\}$ has a finite number of elements.

(2) If $F : I \to \mathbf{R}$ *is increasing, prove that there exists exactly one measure* $\mu \geq 0$ *on* $(I, \mathcal{B})$ *such that*

$$F(y) - F(x) = \mu([x,y])$$

for all x, y *such that* $[x,y] \subset I$ *and* x, $y \notin D_F$.

Prove that $\mu(\{a\}) = F(a+0) - F(a-0)$ *for every* a *in* I.
METHOD. *Uniqueness:* Use the fact (II-3.2) that a Borel measure which is locally finite on an interval is regular, and hence determined by its values on open sets.
Existence: Imitate the construction of the Riemann integral. For every continuous function f with support contained in I, define the integral $\int f d\mu$ as the limit of integrals of step functions

$$\sum_i g(x_i)(F(x_i) - F(x_{i-1})).$$

(3) Let μ be a locally finite nonnegative measure on $(I, \mathcal{B})$ and let $x_0 \in I$. Set $F(x) = \mu([x_0, x))$ if $x > x_0$ and $F(x) = -\mu([x, x_0))$ if $x \leq x_0$. Show that F is increasing and that $F(y) - F(x) = \mu([x, y])$ if $y \notin D_F$.
(4) Let a relation on the set of increasing functions on I be defined as follows: $F_1 \sim F_2$ if there exists a finite or countable subset $D_{1,2}$ of I such that $F_1(y) - F_1(x) = F_2(y) - F_2(x)$ for all $x, y \in I \setminus D_{1,2}$. Show that this defines an equivalence relation on the set of increasing functions on I. Characterize the equivalence classes in terms of measures.

SOLUTION. (1) The number of elements of $D(n; [a, b])$ is bounded above by $[F(b+0) - F(a-0)]n$. Indeed, let $N+1$ be the first integer which is strictly greater than this real number. If $x_0, x_1, \ldots, x_N$ were distinct points in $D(n; [a, b])$, we would have

$$F(b+0) - F(a-0) \geq \sum_{j=0}^{N} [F(x_j + 0) - F(x_j - 0)] \geq \frac{(N+1)}{n},$$

giving a contradiction. Next, let $[a_k, b_k]$ be such that $\cup_{k=1}^{\infty} [a_k, b_k] = I$ with $[a_k, b_k] \subset [a_{k+1}, b_{k+1}]$ for every k.

Since $D_F = \cup_{n=1}^{\infty} \cup_{k=1}^{\infty} D(n; [a_k, b_k])$, we have proved (1).
(2) *Uniqueness.* If μ_1 and μ_2 are two measures with the property described in the problem, we show that $\mu_1((a, b)) = \mu_2((a, b))$ for $[a, b] \subset I$. There exist sequences $\{a_k\}$ and $\{b_k\}$ in $I \setminus D_F$ such that $a_k \downarrow a$ (strictly) and $b_k \uparrow b$ (strictly), since D_F is at most countable. Since $\cup_{k=1}^{\infty} [a, b] = (a, b)$,

$$\mu_1((a, b)) = \lim_{k \to +\infty} \mu_1([a_k, b_k]) = \lim_{k \to +\infty} \mu_2([a_k, b_k]) = \mu_2((a, b)).$$

Since every open set in I is a finite or countable union of pairwise-disjoint open intervals, we conclude that μ_1 and μ_2 coincide on the open subsets of I which are relatively compact in I. It follows that μ_1 and μ_2 are locally finite, since every compact subset of I is contained in an interval $[x, y] \subset I$ with $x, y \in I \setminus D_F$ (so $\mu_1([x, y]) < \infty$ and $\mu_2([x, y]) < \infty$). Hence μ_1 and μ_2 are regular, and since they coincide on the open sets they coincide on the Borel sets.
Existence. Let f be continuous, with compact support contained in $[a, b] \subset I$. Let V_ϵ be the set of step functions g on $[a, b]$ such that $\sup_{x \in [a,b]} |f(x) - g(x)| \leq \epsilon$. (A function g is a step function on $[a, b]$ if there exists a subset $T = \{x_1, \ldots, x_n\}$ of (a, b), with $x_0 = a < x_1 < \ldots < x_n < x_{n+1} = b$, such that g is constant in each interval (x_{i-1}, x_i) $(i = 0, \ldots, n)$.) For a step function g, we define

$$S(g) = \sum_{i=1}^{n} g\left[\frac{x_{i-1} + x_i}{2}\right] [F(x_i - 0) - F(x_{i-1} + 0)],$$

and it is easy to see that, if $g_1 \in V_{\epsilon_1}$ and $g_2 \in V_{\epsilon_2}$,

$$|S(g_1 - g_2)| \leq [F(b) - F(a)](\epsilon_1 + \epsilon_2).$$

This implies that if $g_n \in V_{\frac{1}{n}}$, then $\{S(g_n)\}_{n=1}^{\infty}$ is a Cauchy sequence whose limit $S(f)$ depends only on f and not on the particular sequence g_n chosen.

If g_1 and g_2 are step functions and λ and μ are real numbers, it is easy to show that $S(\lambda g_1+\mu g_2) = \lambda S(g_1)+\mu S(g_2)$. If $g \geq 0$, then $S(g) \geq 0$ trivially. Hence, if f_1 and f_2 are continuous with compact support in $[a,b]$, then $S(\lambda f_1 + \mu f_2) = \lambda S(f_1) + \mu S(f_2)$; if f is nonnegative and continuous, then $S(f) \geq 0$. $S(f)$ is thus a positive linear functional on the space $C_K(I)$ of continuous functions with compact support, and it follows from the Radon-Riesz theorem that there exists a locally finite measure $\mu \geq 0$ on $(I, \mathcal{B})$ such that $\int_I f(x)d\mu(x) = S(f)$.

Next, we show that $\mu([x,y]) = F(y) - F(x)$ if x, $y \notin D_F$ and $[x,y] \subset I$. It suffices to consider the continuous function f_n, defined by $f_n(t) = 0$ if $t \notin [x,y]$ and $f_n(t) = 1$ if $t \in [x+\frac{1}{n}, y-\frac{1}{n}]$, such that f_n is linear on $[x, x+\frac{1}{n}]$ and $[y-\frac{1}{n}, y]$. Then $S(f_n) \to \mu([x,y])$ as $n \to \infty$, while $S(f_n) = F(y-\frac{1}{n}) - F(x+\frac{1}{n}) + r_n + s_n$, with

$$r_n = \int_{[x,x+\frac{1}{n}]} n(t-x)\mu(dt) \quad \text{and} \quad s_n = \int_{[y-\frac{1}{n},y]} n(y-t)\mu(dt).$$

$r_n \leq \mu(\ x, x+\frac{1}{n}]\ ) \to 0$ as $n \to 0$, since $\cap_n (x, x+\frac{1}{n}] = \emptyset$. Similarly, $s_n \to 0$ as $n \to 0$, and the existence of μ with the desired property is proved.

In order to see that $\mu(\{a\}) = F(a+0) - F(a-0)$, it suffices to find $[x_k, y_k] \subset I$ with x_k and $y_k \notin D_f$, $x_k \uparrow a$, and $y_k \downarrow a$. Then, since $\cap_{k=1}^{\infty}[x_k, y_k] = \{a\}$,

$$\mu(\{a\}) = \lim_{k\to 0} \mu([x_k, y_k]) = \lim_{k\to\infty} F(y_k) - F(x_k) = F(a+0) - F(a-0).$$

(3) It is trivial that F is increasing. If $x < y$, it is also clear that $F(y) - F(x) = \mu([x,y))$. Hence, if $y \notin D_f$, there exists a sequence $\{y_k\}$ with $y_k \notin D_f$ such that $y_k \downarrow y$ strictly, and hence $[x,y] = \cap_{k=1}^{\infty}[x, y_k)$. It follows that $\mu([x,y]) = \lim_{k\to\infty} \mu([x, y_k)) = F(y+0) - F(x) = F(y) - F(x)$.
(4) All that is left to check is transitivity. If $F_1 \sim F_2$ and $F_2 \sim F_3$, there exist finite or countable sets $D_{1,2}$ and $D_{2,3}$ such that

$$\begin{aligned} F_1(y) - F_1(x) &= F_2(y) - F_2(x) \quad && \text{if } x \text{ and } y \in I \setminus D_{1,2} \\ F_2(y) - F_2(x) &= F_3(y) - F_3(x) \quad && \text{if } x \text{ and } y \in I \setminus D_{2,3}. \end{aligned}$$

It suffices to introduce $D_{2,3} = D_{1,2} \cup D_{2,3}$ to show that $F_1 \sim F_3$.

The equivalence classes can now be described by using parts (2) and (3) of the problem. They show that there is a bijection between the quotient space and the set of locally finite positive measures on $(I, \mathcal{B})$, i.e. the set of positive Radon measures on I.

REMARKS. 1. Since perhaps as many as 90 per cent of the measures used in practice are measures on $\mathbf{R}$, a description of all the Radon measures ≥ 0

on an open interval is important. Historically, the first measures ≥ 0 were considered by Stieltjes, precisely by means of increasing functions.
2. With every increasing function F on an open interval I, we can thus associate a measure $\mu(dx)$, which is often written $dF(x)$ or $F(dx)$. Conversely, given a measure $\mu \geq 0$ on I, an increasing function F satisfying the hypotheses of (2) is called a *distribution function* for μ. As we have seen, a distribution function for μ is not unique; we can modify (slightly) its value at points of discontinuity (the atoms of μ) and add an arbitrary constant. When μ is a probability measure on $\mathbf{R}$, there are three traditional choices for distribution functions:

$$F_1(x) = \mu((-\infty, x)), \quad F_2(x) = \mu((-\infty, x]), \quad \text{and } F_3(x) = \frac{1}{2}[F_1(x)+F_2(x)].$$

The third appears in the inversion formula for a characteristic function.
3. If we consider a measure $\mu \geq 0$ on a closed interval of the form $(-\infty, b]$, $[a, +\infty)$, or $[a, b]$, we can define its distribution function as above. However, two measures can then have the same distribution function but different masses at the endpoints of the interval.
4. Many identities and inequalities use increasing functions on an interval. It is essential to express the latter in terms of measures in order to understand the former; this also gives a systematic method of proof, although not necessarily the shortest.

Problem II-2. *Specify for which measure on the open interval I each of the following increasing functions is the distribution function (see Problem II-1).*

(1) $I = \mathbf{R}$
   a) $F(x) = x$ b) $F(x) = [x]$ c) $F(x) = \frac{1}{\pi}$ arctan x
(2) $I = (-1, +1)$
   a) $F(x) = \tan \frac{\pi x}{2}$ b) $F(x) = (\operatorname{sign} x)|x|^{\frac{1}{2}}$ c) $F(x) = \frac{1}{\pi} \arcsin x$
(3) $I = (0, +\infty)$
   a) $F(x) = \log x$ b) $F(x) = -[\frac{1}{x}]$ c) $F(x) = (x-1)^+$

(Notation: $[a] = \sup\{n : n \in \mathbf{Z}.\ n \leq a\}$. $a^+ = \sup\{0, a\}$. *sign* $a = +1$ *if* $a > 0$, *sign* $0 = 0$, *and sign* $a = -1$ *if* $a < 0$.*)*

SOLUTION. It is clear that if the function $F'(t)$ exists and is continuous in $[x, y] \subset I$, then the restriction of $\mu(dt) = dF(t)$ to $[x, y]$ is $F'(t)dt$.
(1a) $dF =$ Lebesgue measure on $\mathbf{R}$
  (b)$dF =$ the Dirac measure which is 1 at every integer
  (c) $dF = \frac{1}{\pi}\frac{dx}{1+x^2}$ (Cauchy distribution).
(2a) $dF(x) = \frac{\pi}{2}(1 + \tan^2 \frac{\pi x}{2})dx$ (an unbounded measure on the interval $(-1, +1)$).

(b) $dF(x) = \frac{dx}{2|x|^{\frac{1}{2}}}$ (a bounded measure on $(-1, +1)$ which has unbounded density)

(c) $dF(x) = \frac{dx}{\pi\sqrt{1-x^2}}$ (arcsin distribution)

(3a) $dF(x) = \frac{dx}{x}$ (an unbounded measure)

(b) dF = the Dirac measure which is 1 at every number of the form $\frac{1}{n}$, n a positive integer

(c)$dF(x) = \mathbf{1}_{[1,+\infty)}(x)dx$.

Problem II-3. *Let I be an open interval in* $\mathbf{R}$. *A function G is called convex if its right derivative* $\lim_{\epsilon\downarrow 0}[G(x+\epsilon) - G(x)] = G'_+(x)$ *exists for every x in I and the function $x \mapsto G'_+(x)$ is increasing. (See I-9.2.1.)*

Prove that G is convex if and only if there exists an increasing function F on I such that, for every x_0 in I,

$$G(x) - G(x_0) = \int_{x_0}^{x} F(t)dt.$$

METHOD. For one direction, show that $G'_+(x) = \lim_{\epsilon\downarrow 0} F(x+\epsilon)$. For the other, consider $H(x) = \int_{x_0}^{x} G'_+(x)dt$ and use without proof the fact that, if a function has a right derivative which is zero in an open interval I, it is constant in I.

SOLUTION. Let $G(x) - G(x_0) = \int_{x_0}^{x} F(t)dt$. Then, for $\epsilon > 0$,

$$\frac{1}{\epsilon}[G(x+\epsilon) - G(x)] = \frac{1}{\epsilon}\int_{x}^{x+\epsilon} F(t)dt = \frac{1}{\epsilon}\lim_{\eta\downarrow 0}\int_{x+\eta}^{x+\epsilon} F(t)dt$$

since the function $y \mapsto \int_{y}^{x+\epsilon} F(t)dt$ is continuous on I. Hence, since F is increasing,

$$\lim_{\eta\downarrow 0} F(x+\eta) \leq \frac{1}{\epsilon}[G(x+\epsilon) - G(x)] \leq F(x+\epsilon),$$

which implies that $G'_+(x) = \lim_{\epsilon\downarrow 0} F(x+\epsilon)$. Thus $G'_+(x)$ exists, and it is trivial that it is increasing.

Conversely, if G'_+ is increasing, the function $H(x) = \int_{x_0}^{x} G'_+(t)dt$ satisfies $H'_d(x) = \lim_{\epsilon\downarrow 0} G'_+(x+\epsilon)$, by the first part of the proof. We show that

$$\lim_{\epsilon\downarrow 0} G'_+(x+\epsilon) = G'_+(x).$$

If this were not true, the fact that G'_+ is increasing would imply, for fixed x, the existence of a number k such that $G'_+(x_1) - k < 0 < \lim_{\epsilon\downarrow 0} G'_+(x_1 + \epsilon) - k$. The function $x \mapsto G(x) - kx - G(x_1)$ on $[x_1, x_1 + \epsilon]$ would then attain its minimum at a point $m_\epsilon \in (x_1, x_1+\epsilon]$. But the function G is differentiable at the points of continuity of $G'_+(x)$. Since there exists a sequence $x_0 + \epsilon_n$ of points of continuity of $G'_+(x)$ with $\epsilon_n \downarrow 0$, we can assert that $G'_+(m_{\epsilon_n}) - k = 0$ and hence that $\lim_{n\to\infty} G'_+(x_1+\epsilon_n) = k$. But this contradicts the definition of k.

REMARK. It can be shown that the definition of convex functions given here is equivalent to the following property:

$$G[\lambda x + (1-\lambda)y] \le \lambda G(x) + (1-\lambda)G(y) \quad \text{if } x,\ y \in I \text{ and } \lambda \in [0,1].$$

For a proof of this equivalence and further details of convex functions, the reader may consult Artin[1] or Zygmund[2].

Problem II-4. *Let I be an* open *interval in* $\mathbf{R}$. *Recall (see Problem II-3) that a function $G : I \to \mathbf{R}$ is called* convex *if there exists an increasing function F on I such that, for every x_0 in I,*

$$G(x) - G(x_0) = \int_{x_0}^{x} F(t)dt.$$

If μ is the measure on I given by the distribution function F (see Problem II-2), prove the following assertions.
(1) If $x_0 \le x$, with x and x_0 in I, then

$$\begin{aligned} G(x) - G(x_0) &= (x-x_0)F(x_0+0) + \int_I \mathbf{1}_{(x_0,x]}(u)(x-u)\mu(du) \\ &= (x-x_0)F(x_0-0) + \int_I \mathbf{1}_{[x_0,x]}(u)(x-u)\mu(du). \end{aligned}$$

(2) If $x_0 \ge x$, with x_0 and x in I, then

$$\begin{aligned} G(x) - G(x_0) &= (x-x_0)F(x_0+0) - \int_I \mathbf{1}_{[x,x_0]}(u)(x-u)\mu(du) \\ &= (x-x_0)F(x_0-0) - \int_I \mathbf{1}_{[x,x_0)}(u)(x-u)\mu(du). \end{aligned}$$

SOLUTION. Assume that $x \ge x_0$. If D_F is the set of points of discontinuity of F, then

$$F(t) = F(x_0+0) + \int_I \mathbf{1}_{(x_0,t]}(u)\mu(du) \quad \text{if } t > x_0 \text{ and } t \notin D_F.$$

---

[1] E. Artin, *The Gamma Function* (New York: Holt, Rinehart and Winston 1964), 1–6.

[2] A. Zygmund, *Trigonometric Series* (Cambridge: Cambridge University Press 1959), 21–26.

Since D_F is finite or countable, we can thus write

$$\int_{x_0}^{x} F(t)dt = (x - x_0)F(x_0 + 0) + \int_{x_0}^{x} dt \int_I \mathbf{1}_{(x_0,t]}(u)\mu(du).$$

Applying Fubini's theorem (I-8.5) to the last integral gives

$$\int_{x_0}^{x} dt \int_I \mathbf{1}_{(x_0,t]}(u)\mu(du) = \int_I \mu(du) \int_{x_0}^{x} \mathbf{1}_{(x_0,t]}(u)dt = \int_I f(u)\mu(du),$$

where $f(u) = \int_I \mathbf{1}_{(x_0,t]}(u)dt$. Calculating $f(u)$ shows that $f(u) = x - u$ if $x_0 < u \leq x$ and $f(u) = 0$ otherwise, completing the proof of the first formula. The proofs of the other three are similar.

REMARKS. If μ has no atoms and $x_0 \leq x$, we can replace the notation $\int_I \mathbf{1}_{(x_0,x]}(u)g(u)du = \int_I \mathbf{1}_{[x_0,x]}(u)g(u)du$ by $\int_{x_0}^{x} g(u)\mu(du)$, since the latter is unambiguous in this case. If $x \leq x_0$, we write $\int_{x_0}^{x} g(u)\mu(du) = -\int_I \mathbf{1}_{[x,x_0]}(u)\mu(du)$, which permits us to state the relation of Chasles: $\int_a^c = \int_a^b + \int_b^c$ for arbitrary a, b, and c in I. However, this relation does not hold if μ has atoms.

Problem II-5. *Let M_1 be the set of measures $\mu \geq 0$ on $(0, +\infty)$ equipped with its Borel algebra, such that $\int_0^{\infty} \mathbf{1}_{[x,+\infty)}(u)u\mu(du) < \infty$ for every $x > 0$. (1) Let G be a convex function on $(0, +\infty)$ (see Problem II-4) such that* $\lim_{x\to+\infty} G(x) = 0$. *Prove that there exists a unique μ in M_1 such that*

$$(i) \qquad G(x) = \int_0^{+\infty} (u - x)^+\mu(du) \quad \textit{for every } x > 0,$$

where $a^+ = \max(0, a)$, and that $\int_0^{+\infty} u\mu(du) = \lim_{x\to 0} G(x) \leq +\infty$.
(2) Conversely, let $\mu \in M_1$. Show that (i) defines a convex function G on $(0, +\infty)$ *such that* $\lim_{x\to+\infty} G(x) = 0$.

METHOD. Let $F(x)$ be as in Problem II-4 and show that $F(x) \leq 0$ and that $\lim_{x\to+\infty} xF(x) = 0$. Then use Problem II-4.

SOLUTION. (1) If there exists $x_0 > 0$ such that $F(x_0) > 0$, then, for $x > x_0$,

$$G(x) \geq G(x_0) + (x - x_0)F(x_0),$$

which implies that $\lim_{x\to+\infty} G(x) = +\infty$. But this is a contradiction. Therefore

$$G(x) - G(\frac{x}{2}) = \int_{\frac{x}{2}}^{x} F(t)dt \leq \frac{x}{2}F(x) \leq 0.$$

Hence $\lim_{x\to+\infty} xF(x) = 0$. Applying (2) of Problem II-4 and letting $x_0 \to +\infty$,

$$G(x) = \int_0^{+\infty} \mathbf{1}_{[x,+\infty)}(u)(u-x)\mu(du).$$

This proves formula (i), with $G(x) \geq 0$ and $\int_0^\infty \mathbf{1}_{[x,+\infty)}(u)u\mu(du) < \infty$ (and thus $\mu \in M_1$).

To prove uniqueness, let $\mu_1 \in M_1$ such that $G(x) = \int_0^\infty (u-x)^+\mu_1(du)$. Then

$$\begin{aligned} -\int_x^{+\infty} F(t)dt = G(x) &= \int_0^{+\infty} (u-x)^+\mu_1(du) \\ &= \int_0^{+\infty} \mu_1(du) \int_0^{+\infty} \mathbf{1}_{[x,u]}(t)dt \\ &= \int_x^{+\infty} dt \int_0^{+\infty} \mathbf{1}_{[t,+\infty)}(u)\mu_1(du), \end{aligned}$$

using Fubini's theorem for the last equality. Setting $F_1(t) = -\mu_1(\,[t,+\infty)\,)$, we then have $\int_x^\infty F(t)dt = \int_x^\infty F_1(t)dt$ for all $x > 0$. Next, considering the two nonnegative measures $-F(t)dt$ and $-F_1(t)dt$ on $(0,+\infty)$, we see that they have the same distribution function $-G(x)$, and hence are equal. It follows that F and F_1 coincide almost everywhere on $(0,+\infty)$. Since they are increasing functions, they agree everywhere outside D_F, which shows that $\mu = \mu_1$.

To prove the last formula of (1), observe that, for fixed $u > 0$, $(u-x)^+$ increases to u as x decreases to zero; the conclusion follows from the monotone convergence theorem.

(2) If G is defined by formula (i), then the fact that $(u-x)^+ \downarrow 0$ as $x \to \infty$ implies, by monotone convergence, that $\lim_{x\to+\infty} G(x) = 0$. In order to show that G is convex, we reverse the argument of (1), expressing F in terms of μ and showing that $G(x) - G(x_0) = \int_{x_0}^x F(t)dt$.

REMARKS.

1. The proof of (2) can be shortened by using the characterization of convex functions mentioned in Problem II-4.
2. The measure $x\mu(dx)$ is not necessarily bounded: $G(x) = \frac{1}{x}$ gives $\mu(dx) = \frac{dx}{x^3}$.

Problem II-6. *Let M be the set of measures $\nu \geq 0$ on $(0,+\infty)$ equipped with its Borel algebra, such that $\nu([x,+\infty)) < +\infty$ for every $x > 0$. If k is a positive integer, we denote by C_k the set of functions g on $(0,+\infty)$ such that $G(x) = (-1)^{k-1}g^{(k-1)}(x)$ exists and is convex and also that $\lim_{x\to+\infty} g(x) = \lim_{x\to+\infty} G(x) = 0$.*

(1) If $g \in C_k$, show that there exists a unique ν in M such that

$$(i) \qquad g(x) = \int_0^\infty \left[\left(1 - \frac{x}{u}\right)^+\right]^k \nu(du) \quad \text{for every } x > 0.$$

(2) Conversely, let $\nu \in M$. Show that (i) defines an element of C_k.

METHOD. (1) First use Taylor's formula to show that $\lim_{x\to+\infty} g^{(j)}(x) = 0$ for $j = 0, 1, \ldots, k-1$, then use Problem II-5.

SOLUTION. (1) If $g \in C_k$, then by Taylor's formula there exists a number $\theta(x, h)$ in $(0, 1)$ such that, for x, $x + h > 0$,

$$(ii) \qquad g(x+h) - \frac{h^{k-1}}{(k-1)!} g^{(k-1)}[x + \theta(x,h)h] = \sum_{j=0}^{k-2} \frac{h^j}{j!} g^{(j)}(x).$$

We set $P_j(h, x) = g^{(j)}(x) + \sum_{i=1}^{k-j-1} h^i \frac{g^{(j+i)}(x)}{(j+1)!}$ for $j = 0, 1, \ldots, k-2$ and show by induction on j that $\lim_{x\to+\infty} P_j(h, x) = 0$. Since $P_0(h, x)$ is the right-hand side of (ii), the fact that g is in C_k implies that $\lim_{x\to+\infty} P_0(h, x) = 0$. Suppose that $\lim_{x\to+\infty} P_j(h, x) = 0$ for some $j < k-2$. Then $P_j(0, x) = g^{(j)}(x) \to 0$ as $x \to +\infty$; hence $P_{j+1}(h, x) = \frac{1}{h}[P_j(h, x) - g^{(j)}(x)] \to 0$ as $x \to +\infty$, which proves the induction step and shows that $\lim_{x\to+\infty} g^{(j)}(x) = 0$ for $j = 0, 1, \ldots, k-1$.

By Problem II-5, there exists a unique measure $\mu \geq 0$ on $(0, +\infty)$ such that $\int_0^\infty \mathbf{1}_{[x,+\infty]}(u) u \mu(du) < \infty$ for every $x > 0$ and

$$(iii) \qquad (-1)^{k-1} g^{(k-1)}(x) = \int_0^\infty (u-x)^+ \mu(du) \quad \text{for all } x > 0.$$

We now show by induction on j that

$$(iv) \qquad (-1)^{k-j} g^{(k-j)}(x) = \int_0^\infty \left[(u-x)^+\right]^j \mu(du) \quad \text{for all } x > 0.$$

and that

$$(v) \qquad \int_0^\infty \mathbf{1}_{[x,+\infty)}(u) u^j \mu(du) < \infty.$$

Clearly (iv) and (v) hold for $j = 1$. Assume that they hold for $j < k$; then, for $0 < x < x_0$,

$$(-1)^{k-j-1} \left[g^{(k-j-1)}(x) - g^{(k-j-1)}(x_0)\right] = (-1)^{k-j} \int_x^{x_0} g^{(k-j)}(t) dt.$$

Since $\lim_{x_0\to+\infty} g^{(k-j-1)}(x_0) = 0$,

$$\begin{aligned}
(-1)^{k-j-1} g^{(k-j-1)}(x) &= (-1)^{k-j} \int_x^\infty g^{(k-j)}(t) dt \\
&= \int_x^\infty \frac{dt}{j!} \int_0^\infty [(u-t)^+]^j \mu(du) \\
&= \frac{1}{(j+1)!} \int_0^\infty [(u-x)^+]^{j+1} \mu(du)
\end{aligned}$$

The last equality was obtained by applying Fubini's theorem to the non-negative function

$$f(u,t) = 1 \text{ if } u > t \geq x, \quad f(u,t) = 0 \text{ otherwise,}$$

and to the measure $dt\mu(du)$ on $([x,+\infty))^2$. Since $g^{(k-j-1)}(x)$ is finite, $\int_x^\infty u^{j+1}\mu(du)$ exists; thus (v) holds for all j by induction. In particular, it follows that

$$g(x) = \frac{1}{k!}\int_0^\infty [(u-x)^+]^k \mu(du).$$

We now define a measure ν in M by $d\nu(u) = \frac{u^k}{k!}d\mu(u)$; that is, $\nu(B) = \int_B \frac{u^k}{k!}d\mu(u)$ for every Borel subset of $(0,+\infty)$. Then $\nu(\,[x,+\infty)\,) < \infty$ for every $x > 0$, and formula (i) also holds.

Uniqueness is easy: if $g(x) = \int_0^{+\infty} \left[(1-\frac{x}{u})^+\right]^k \nu_1(du)$, we set $\mu_1(du) = \frac{k!}{u^k}\nu_1(du)$. Clearly

$$G(x) = \int_0^{+\infty} (u-x)^+\mu(du) = \int_0^{+\infty} (u-x)^+\mu_1(du).$$

Problem II-5 gives the desired uniqueness.

(2) If g is defined by formula (i), we set $\mu(du) = \frac{k!}{u^k}\nu(du)$, and it follows from (i) that (vi) holds and that $G(x) = (-1)^{k-1}g^{k-1}(x) = \int_0^\infty (u-x)^+\mu(du)$. Since $(u-x)^+ \downarrow 0$ as $x \to +\infty$ for fixed x, the monotone convergence theorem implies that $\lim_{x\to+\infty} g(x) = \lim_{x\to+\infty} G(x) = 0$. The convexity of G was seen in Problem II-5.

REMARK. It is clear that the functions $f_u(x) = \left[(1-\frac{x}{u})^+\right]^k$ play the role of extremals in C_k; formula (i) shows that the functions in C_k are "barycenters" of the f_u. Formula (i) plays a role in the probability distributions of Polya and Askey. (See Problem III-5.)

Problem II-7. *Let u be a decreasing function defined on $(0,+\infty)$ such that $u \to 0$ as $x \to \infty$ and $\int_0^\infty x^2u(x)dx < \infty$. Show that, for every $y > 0$,*

$$y^2\int_y^{+\infty} u(x)dx \leq \frac{4}{9}\int_0^{+\infty} x^2u(x)dx \quad \textit{(K.F.Gauss).}$$

Describe in detail the case of equality.

METHOD. Consider a measure μ on $(0,+\infty)$ for which $-u$ is a distribution function.

SOLUTION. Proceeding as in Problem II-5, we see easily that

$$\int_y^{+\infty} u(x)dx = \int_0^\infty (x-y)^+\mu(dx) \quad \text{and} \quad \int_0^\infty x^2u(x)dx = \int_0^\infty \frac{x^3}{3}\mu(dx).$$

The assertion is thus equivalent to $y^2(x-y)^+ \leq \frac{4}{27}x^3$ for all $x,\ y > 0$. In other words, considering $f(h) = \frac{4}{27}(1+h)^3 - h^+$ defined on $(-1,+\infty)$, we must show that $f(\frac{x}{y}-1) \geq 0$ for all x and $y > 0$.

It thus suffices to consider $f(h)$ on $(-1,+\infty)$. It is clear that $f(h) > 0$ if $h \leq 0$. Moreover, $f'(h) = \frac{4}{9}(1+h)^2 - 1$ is zero in $(0,+\infty)$ only for $h = \frac{1}{2}$. It follows easily that $f(h) \geq 0$ for $h > -1$, and that $h = \frac{1}{2}$ is the only solution in $(-1,+\infty)$ of $f(h) = 0$. Hence

$$\frac{4}{9}\int_0^\infty x^2 u(x)dx - y^2\int_y^\infty u(x)dx = y^3\int_0^\infty f\left(\frac{x}{y}-1\right)\mu(dx) \geq 0.$$

Moreover, for fixed $y > 0$, this is zero if $x \mapsto f(\frac{x}{y}-1)$ is zero μ-almost everywhere; that is, if μ is proportional to a Dirac measure concentrated at $\frac{3y}{2}$, and thus if $u = k\mathbf{1}_{(0,\frac{3y}{2}]}$.

Problem II-8. *Let u be a decreasing function defined on $(-a,+\infty)$, with $a > 0$, such that $u \to 0$ as $x \to \infty$ and $\int_{-a}^{+\infty} u(x)dx < \infty$. Show that*

$$\int_0^y u(x)dx \leq \frac{y}{y+a}\int_{-a}^{+\infty} u(x)dx \quad \forall y > 0,$$

and describe in detail the case of equality.

METHOD. Consider a measure μ on $(-a,+\infty)$ for which $-u$ is a distribution function.

SOLUTION. Proceeding as in Problem II-5, we easily obtain

$$\int_0^y u(x)dx = \int_{-a}^{+\infty} \mathbf{1}_{[0,\infty)}(x)\min(x,y)\mu(dx)$$

and

$$\frac{y}{y+a}\int_{-a}^{+\infty} u(x)dx = \int_{-a}^{+\infty} \frac{y(x+a)}{y+a}\mu(dx).$$

The inequality $\mathbf{1}_{[0,+\infty)}(x)\min(x,y) \leq \frac{y(x+a)}{y+a}$ for $x > -a$ can be verified directly, and equality can occur only if μ is proportional to a Dirac measure concentrated at y.

Problem II-9. *Let F be an increasing function on $[a,b]$ and let f be an integrable function on $[a,b]$. Show that there exists a number ξ in $[a,b]$ such that*

$$\int_a^b f(x)F(x)dx = F(a)\int_a^\xi f(x)dx + F(b)\int_\xi^b f(x)dx.$$

(Second mean value theorem for integrals)

METHOD. Show that this can be reduced to the case where $F(a) = 0$ and $F(b) = 1$, and consider a probability measure μ on $[a, b]$ such that $F(x) = \mu([a, x])$ for $x \notin D_F = \{x : a < x < b \text{ and } F(x-0) < F(x+0)\}$.

SOLUTION. We first prove the formula with the hypothesis that $F(a) = 0$ and $F(b) = 1$. In this case,

$$\int_a^b f(x)F(x)dx = \int_a^b f(x)dx \int_a^b \mathbf{1}_{[a,x]}(t)\mu(dt) = \int_a^b \mu(dt) \int_t^b f(x)dx.$$

The first equality holds because $F(x) = \mu([a, x])$ except on a set which is at most countable (see Problem II-4), and the second by Fubini's theorem. We next set $f_1(t) = \int_t^b f(x)dx$. Since this is a continuous function on $[a, b]$, there exist α and β in $[a, b]$ such that $f_1(\alpha) \leq f_1(t) \leq f_1(\beta)$ for every t; since μ is a probability measure,

$$f_1(\alpha) \leq \int_a^b f_1(t)\mu(dt) \leq f_2(\beta).$$

The intermediate value theorem guarantees the existence of ξ in $[a, b]$ such that $\int_a^b f_1(t)\mu(dt) = f_1(\xi)$, and this proves the formula.

Finally, to reduce the proof to the case where $F(a) = 0$ and $F(b) = 1$, consider $F_1(x) = \frac{F(x)-F(a)}{F(b)-F(a)}$, the case $F(b) = F(a)$ being trivial.

Problem II-10. *Let μ be a probability measure on $[0, 1]$. Set $m = \int_0^1 x\mu(dx)$ and $\sigma^2 = \int_0^1 x^2\mu(dx) - m^2$. Show that $\sigma^2 \leq \frac{1}{4}$. Describe in detail the case of equality.*

SOLUTION.

$$\frac{1}{4} - \sigma^2 = \frac{1}{4} + m^2 - \int_0^1 x^2\mu(dx) = (\frac{1}{2} - m)^2 + \int_0^1 x(1-x)\mu(dx) \geq 0.$$

$\sigma^2 = \frac{1}{4}$ implies that $x(1-x) = 0$ μ-almost everywhere on $[0, 1]$ and that $m = \frac{1}{2}$. It follows that $\mu = p\delta_0 + (1-p)\delta_1$ with $0 \leq p \leq 1$. The condition $m = \frac{1}{2}$ implies that $p = \frac{1}{2}$.

Problem II-11. *Let f be a positive decreasing function on $(0, 1]$ such that $\int_0^1 f(x)dx = 1$ and let $\lambda \in [0, 1]$. Let $P(dx) = \lambda\delta_0(dx) + (1-\lambda)f(x)dx$, where δ_0 is the Dirac measure at the origin, let $m(\lambda, f) = \int_0^1 xP(dx)$, and let $\sigma^2(\lambda, f) = \int_0^1 x^2P(dx) - m^2(\lambda, f)$.*
(1) Show that $\sigma^2(\lambda, f) \leq \frac{1}{9}$. Describe in detail the case of equality.
(2) Show that $\sigma^2(0, f) < \frac{1}{9}$. Is this inequality the best possible?

METHOD. If D_f is the set of points of discontinuity of f in $(0,1]$, consider the measure ν on $(0,1]$ such that $f(x) = \nu([x,1])$ if $x \notin D_f$ and show that $\mu(dt) = t\nu(dt)$ is a probability measure on $(0,1]$.

SOLUTION. (1)

$$1 = \int_0^1 f(x)dx = \int_0^1 dx \int_0^1 \mathbf{1}_{[x,1]}(1)\nu(dt) = \int_0^1 \nu(dt) \int_0^t dx = \int_0^1 t\nu(dt).$$

The second equality holds because $f(x) = \nu([x,1])$ outside the set D_f, which has Lebesgue measure zero, and the third by Fubini. Hence μ is a probability measure on $(0,1]$. Similarly,

$$\int_0^1 xf(x)dx = \frac{1}{2}\int_0^1 t\mu(dt) \quad \text{and} \quad \int_0^1 x^2 f(x) = \frac{1}{3}\int_0^1 t^2\mu(dt).$$

Hence

$$\begin{aligned}\frac{1}{9} - \sigma^2(\lambda, f) &= \frac{1}{9} + m^2(\lambda, f) - \frac{1-\lambda}{3}\int_0^t t^2\mu(dt) \\ &= (\frac{1}{3} - m(\lambda, f))^2 + \frac{1-\lambda}{3}\int_0^1 t(1-t)\mu(dt).\end{aligned}$$

Since $t(1-t) \geq 0$ on $(0,1]$, the inequality is proved.

Thus $\sigma^2(\lambda, f) = \frac{1}{9}$ implies that $t(1-t) = 0$ μ-almost everywhere on $(0,1]$ and that $m(\lambda, f) = \frac{1}{3}$. Hence $\mu = \delta_1$, the Dirac measure at 1, and $\lambda = \frac{1}{3}$. The function f is then equal to the constant 1 in $(0,1)$.

(2) We have just seen that

$$\frac{1}{9} - \sigma^2(0, f) = \left[\frac{1}{3} - \int_0^1 t\mu(dt)\right]^2 + \frac{1}{3}\int_0^1 t(1-t)\mu(dt)$$

cannot vanish. However, the inequality given is the best possible. Let $f_\epsilon(x) = \frac{2}{3} + \frac{1}{3\epsilon}$ if $0 < x < \epsilon$ and $f_\epsilon(x) = \frac{2}{3}$ if $\epsilon < x < 1$. Then, with $\mu_2 = \frac{1}{3}\delta_\epsilon + \frac{2}{3}\delta_1$, we obtain $\frac{1}{9} - \sigma^2(0, f_\epsilon) = \frac{1-\epsilon}{3}\epsilon$; this tends to 0 if $\epsilon \to 0$.

REMARK. If G is a convex function from $(0,1)$ to $[0,1)$, it can be shown that the measure P on $[0,1)$ which is the image under G of Lebesgue measure on $(0,1)$ is of the type considered in the problem. Hence

$$\int_0^1 G^2(x)dx = \left[\int_0^1 G(x)dx\right]^2 \leq \frac{1}{9}.$$

Problem II-12. *Let n be a positive integer and let $\alpha, a_1, \ldots, a_n, c_1, \ldots, c_n$ be real numbers such that $a_1 < a_2 < \ldots < a_n$ and $c_j > 0$ for $j = 1, \ldots, n$.*

Let $\overline{\mathbf{C}}$ and $\overline{\mathbf{R}}$ denote the complex and the real numbers completed by a point at infinity ∞. Consider the function $f : \overline{\mathbf{C}} \to \overline{\mathbf{C}}$ defined by $f(x) = \infty$ if $x \in \{\infty, a_1, \ldots, a_n\}$ and

$$f(x) = x + \alpha - \sum_{j=1}^{n} \frac{c_j}{x - a_j} \quad \text{if} \quad x \notin \{\infty, a_1, \ldots, a_n\}.$$

The function $T : \overline{\mathbf{R}} \to \overline{\mathbf{R}}$ is the restriction of f to $\overline{\mathbf{R}}$. Lebesgue measure on $\overline{\mathbf{R}}$ is the measure m such that $m(\{\infty\}) = 0$ and the restriction of m to $\mathbf{R}$ is the usual measure.

(1) Let $y \in \mathbf{R}$. Show that the equation in x given by $f(x) = y$ has exactly $n+1$ real roots $\{x_j(y)\}_{j=0}^n$ such that $a_j < x_j(y) < a_{j+1}$ (with the convention that $a_0 = -\infty$ and $a_{n+1} = +\infty$). Show that $\sum_{j=0}^n x_j'(y) = 1$ and conclude that T preserves m. That is, for every F in $L^1(m)$,

$$\int_{\overline{\mathbf{R}}} F(T(x))m(dx) = \int_{\overline{\mathbf{R}}} F(x)m(dx).$$

(2) Prove by induction on the integer $k \geq 0$ that, for every $z \in \mathbf{C}$,

$$\sum_{j=0}^{n} [x_j(y) - z]^{-k-1} x_j'(y) = \frac{1}{k!}\left(\frac{\partial}{\partial z}\right)^k [y - f(z)]^{-1}.$$

(3) Let g be a nonnegative rational function such that $\int_{\overline{\mathbf{R}}} g(x)m(dx) < \infty$. Prove that there exists a rational function g_1 with the same properties and such that the image $g(x)m(dx)$ under T is $g_1(x)m(dx)$. Conclude from (2) that, if z_1 is a pole of g_1 with multiplicity $m_1 > 0$, there exists a pole z of g with multiplicity m such that $f(z) = z_1$ and $m_1 \leq m$. Calculate g_1 when

$$f(x) = x - \frac{1}{x} \quad \text{and} \quad g(x) = \frac{2x^2}{\pi(x^2+1)^2}.$$

(4) Let $z = a + ib \in \mathbf{C}$, with $b > 0$. The Cauchy measure γ_z on $\overline{\mathbf{R}}$ is defined by $\gamma_z(dx) = \frac{bm(dx)}{\pi[(x-a)^2+b^2]}$. Prove, using (3), that the image of γ_z under T is $\gamma_{f(z)}$.

SOLUTION. (1) We begin by sketching the graph of $T(x)$ for real x. Since

$$T'(x) = 1 + \sum_{j=1}^{n} \frac{c_j}{(x - a_j)^2} > 1,$$

the restriction of T to the interval (a_j, a_{j+1}) for $j = 0, \ldots, n$ is a homeomorphism on $\mathbf{R}$. If $x_j(y)$ is its inverse, it is clear that the $\{x_j(y)\}_{j=0}^n$ are the $n+1$ real roots of the equation $f(x) = y$. Next, $f(x) = y \Leftrightarrow P_y(x) = 0$,

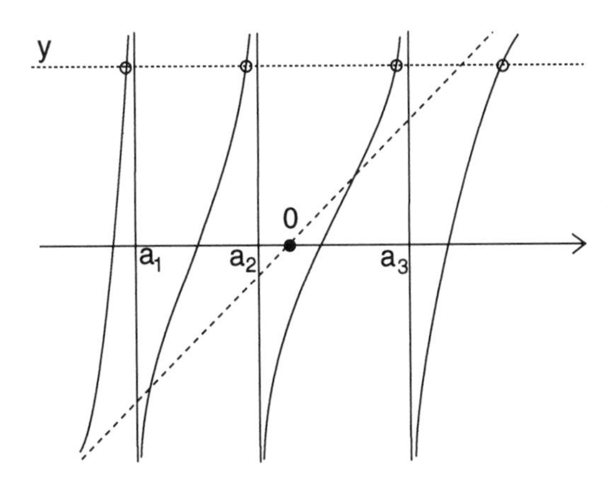

where $P_y(x)$ is defined by

$$P_y(x) = (x + \alpha - y) \prod_{j=1}^{n} (x - a_j) - \sum_{j=1}^{n} c_j \prod_{k \neq j} (x - a_k).$$

Since $\delta^0 P_y = n+1$, there are no other roots in $\mathbf{C} \setminus \mathbf{R}$. The sum of the roots of this polynomial is therefore

$$\sum_{j=0}^{n} x_j(y) = y + \sum_{j=1}^{n} a_j - \alpha,$$

whence $\sum_{j=0}^{n} x_j'(y) = 1$. Thus

$$\begin{aligned} \int_{\mathbf{R}} F(T(x)) m(dx) &= \sum_{j=0}^{n} \int_{a_j}^{a_j+1} F(T(x)) m(dx) \\ &= \sum_{j=0}^{n} \int_{-\infty}^{+\infty} F(y) x_j'(y) dy = \int_{\mathbf{R}} F(y) m(dy). \end{aligned}$$

(2) For $k = 0$, since $\sum_{j=0}^{n} [x_j(y) - z]^{-1} x_j'(y)$ is the logarithmic derivative of $\prod_{j=0}^{n} [x_j(y) - z]$ with respect to y, we must show that this product is a constant multiple (with respect to y) of $y - f(z)$. Since it is the product of the roots of $P_y(x + z)$, a polynomial in x, we have

$$\prod_{j=0}^{n} (x_j(y) - z) = (f(z) - y) \prod_{j=1}^{n} (a_j - z),$$

which settles the case $k = 0$. The general case is handled by differentiating the formula for $k = 0$ with respect to z:

$$\frac{1}{k!}\left(\frac{\partial}{\partial z}\right)^k [x_j(y) - z]^{-1} = [x_j(y) - z]^{-k-1}.$$

(3) If $F \in L^1(g\ dm)$, we proceed as in (1) by writing

$$\int_{\mathbf{R}} F(T(x))g(x)m(dx) = \sum_{j=0}^{n} \int_{-\infty}^{+\infty} F(y)g(x_j(y))x_j'(y)dy,$$

and we set

$$g_1(y) = \sum_{j=0}^{n} g(x_j(y))x_j'(y).$$

We now show that g_1 has the properties claimed. Let $z_1, \ldots, z_N$ be the poles of g, with respective multiplicities $m_1, \ldots, m_N$. Then, by the theorem on partial fraction expansions, there exist polynomials $P_1, P_2, \ldots, P_N$, of degrees $m_1, \ldots, m_N$, such that $P_k(0) = 0$ for $k = 1, \ldots, N$ and

$$g(x) = \sum_{k=1}^{N} P_k\left(\frac{1}{x - z_k}\right).$$

($\int_{\mathbf{R}} g(x)m(dx) < \infty$ implies that the z_k are not real and that the entire part of g is zero.) To use (2), note that there exists a coefficient $f_k(z)$ independent of y such that

$$\frac{1}{k!}\left(\frac{\partial}{\partial z}\right)^k [y - f(z)]^{-1} = [y - f(z)]^{-k-1} f_k(z).$$

It follows that there exist polynomials Q_k of degree $\leq m_k$ such that

$$g_1(z) = \sum_{k=1}^{N} Q_k\left[\frac{1}{y - f(z_k)}\right],$$

which shows that g_1 is a rational function with the stated properties. That $\int g_1(y)m(dy) < \infty$ comes from the fact that the image of a bounded measure is bounded. (See Problem I-14.)

The calculation in the numerical example above gives $g_1(y) = \frac{1}{\pi(y^2+1)}$, which shows that the multiplicities of the poles can in fact decrease in passing from g to g_1.

(4) Let $g(x) = \frac{b}{\pi(x-a)^2+b^2}$. Its poles are $a \pm ib$, with multiplicity 1. The only possible poles of g_1 are thus $f(a \pm ib)$, with multiplicity 0 or 1. But multiplicity 0 is impossible, as this would imply $\int g_1(y)dy = +\infty$. Hence $g_1(y)m(dy) = \gamma_{f(z)}(dy)$.

REMARKS. 1. A Cayley function is a function of the form

$$f(x) = c_0 x + \alpha - \sum_{j=1}^{n} \frac{c_j}{x - a_j},$$

where $c_j \geq 0$, $j = 0, 1, \ldots, n$ and α, $a_1, \ldots, a_n$ are real. If $c_0 = 0$ and $n = 1$, it is a positive linear fractional transformation; that is, $f(x) = \frac{ax+b}{cx+d}$ with a, b, c, and d real and $ad - bc > 0$. It is easy to see that all Cayley functions can be obtained by composing positive linear fractional transformations with the Cayley functions corresponding to $c_0 = 1$.
2. It is easy to see that if f is a positive linear fractional transformation and T is its restriction to $\mathbf{R}$, then the image of γ_z under T is $\gamma_{f(z)}$. This observation, the remark above, and result (4) of the problem show that the property holds for all Cayley functions.
3. Conversely, let $T : \overline{\mathbf{R}} \to \overline{\mathbf{R}}$ be a rational function such that, for every z with positive imaginary part, the image of γ_z under T is a Cauchy distribution γ_{z_1} (where z_1 depends on z). It can be proved that T is the restriction to the real axis of a Cayley function.
4. On the other hand, a Cayley function with $c_0 > 0$ maps Lebesgue measure m to $c_0 m$. If $c_0 = 0$, the image measure is no longer a Radon measure on $\mathbf{R}$. For example, $f(x) = -\frac{1}{x}$ maps $m(dx)$ to $\frac{m(dx)}{x^2}$.

Problem II-13. *The half-plane* $\mathbf{R}_+^2 = \{(x, y) : x \in \mathbf{R} \text{ and } y > 0\}$ *is equipped with the measure* $\mu(dx, dy) = \frac{dx\,dy}{y^2}$. *What is the image* ν *on* $[1, +\infty)$ *of this measure under the mapping* $(x, y) \mapsto v(x, y) = \frac{1}{2y}(1 + x^2 + y^2)$ *(in the sense of Problem I-14)?*

SOLUTION. We observe first that ν exists. If K is a compact subset of $[1, +\infty)$, there exists R such that $[1, R] \supset K$. But $v^{-1}([1, R]) = \{(x, y) \in \mathbf{R}_+^2 : x^2 + (y - R)^2 \leq R^2 - 1\}$ is a compact subset of $\mathbf{R}_+^2$ and hence has finite measure. We now evaluate $F(R) = \mu((v^{-1}([1, R]))) = \displaystyle\int\int_{v^{-1}([1,R])} \frac{dx\,dy}{y^2}$. Since $v^{-1}([1, R])$ is a disk in the Euclidean plane, centered at $(0, R)$ and with radius $\sqrt{R^2 - 1}$, passing to polar coordinates gives

$$F(R) = \int_0^{\sqrt{R^2-1}} \rho d\rho \int_{-\pi}^{+\pi} \frac{d\theta}{(R + \rho \sin\theta)^2} = \int_0^{1-\frac{1}{R^2}} u du \int_{-\pi}^{+\pi} \frac{d\theta}{(1 + u \sin\theta)^2}.$$

It is clear that $F'(R) = \frac{1}{R^3} \int_{-\pi}^{+\pi} \frac{d\theta}{[1 + (1 - \frac{1}{R^2})^{\frac{1}{2}} \sin\theta]^2}$ is a continuous function which will be the density of the measure ν we seek. For simplicity, we set $\cos\alpha = (1 - \frac{1}{R^2})^{\frac{1}{2}}$, with $0 < \alpha \leq \frac{\pi}{2}$. Then, using the change of variables $\theta = 2\arctan t$,

$$F(R) = \sin^3\alpha \int_{-\pi}^{+\pi} \frac{d\theta}{(1 + \cos\alpha \sin\theta)^2} = \sin^3\alpha \int_{-\infty}^{+\infty} \frac{(1 + t^2)dt}{[(t - \cos\alpha)^2 + \sin^2\alpha]^2}.$$

$F'(R) = 2\pi$ by the change of variable $\beta = \arg \frac{t-\cos\alpha}{\sin\alpha}$. Summarizing,

$$\nu(dR) = 2\pi \mathbf{1}_{[1,\infty)}(R)dR.$$

Problem II-14. *Let $\{\mu_n\}_{n\geq 0}$ be a sequence of positive measures on $\mathbf{R}$, each with total mass ≤ 1. Suppose that μ_n converges weakly to μ_0 as $n \to \infty$ and that*

$$M = \sup_n \int_{-\infty}^{+\infty} x^2 \mu_n(dx) < \infty.$$

(1) Show that μ_n converges narrowly to μ_0 as $n \to \infty$.

(2) Show that $\displaystyle\int_{-\infty}^{+\infty} |x|\mu_n(dx) \to \int_{-\infty}^{+\infty} |x|\mu_0(dx)$ as $n \to \infty$.

(3) Show by a counterexample that $\displaystyle\int_{-\infty}^{+\infty} x^2\mu_n(dx)$ does not necessarily tend to $\displaystyle\int_{-\infty}^{+\infty} x^2\mu_0(dx)$.

METHOD. Use Theorem II-6.8.

SOLUTION. (1) Let K_a^c be the complement in $\mathbf{R}$ of $[-a, +a]$. Then

$$M \geq \int_{K_a^c} x^2\mu_n(dx) \geq a^2\mu_n(K_a^c).$$

Thus $\mu_n\left(K^c_{M^{\frac{1}{2}}\epsilon^{-\frac{1}{2}}}\right) \leq \epsilon$ for $\epsilon > 0$, and Theorem II-6.8 gives the result.

(2) Let A be the set of $a > 0$ for which $\mu_0(\{-a, +a\}) = 0$. Then, if $a \in A$, the restriction ν_n of μ_n to the open set $(-a, +a)$ converges narrowly to the restriction ν_0 of μ_0 to $(-a, +a)$. For if $f \in C_0((-a, a))$, then f can be extended by 0 to $\mathbf{R} \setminus (-a, +a)$, and $\int f\mu_n \to \int f\mu_0$ as $n \to \infty$. Hence $\nu_n \to \nu_0$ weakly as $n \to \infty$. Since $a \in A$, for every $\epsilon > 0$ there exists an open set O_ϵ containing $\pm a$ such that $\mu(O_\epsilon) \leq \frac{\epsilon}{4}$.

Let f be a nonnegative continuous function on $\mathbf{R}$ which is compactly supported, equals 1 on O_ϵ, and satisfies $\int f_\epsilon(x)\mu_0(dx) \leq \frac{\epsilon}{2}$. Then $\lim_{n\to\infty} \int f_\epsilon\mu_n = \int f_\epsilon\mu \leq \frac{\epsilon}{2}$; hence, for n sufficiently large, $\mu_n(O_\epsilon) \leq \epsilon$. Theorem II-6.8 implies that ν_n converges narrowly to ν_0. Hence $\int_{-a}^{a} |x|\mu_n(dx) \to \int_{-a}^{a} |x|\mu_0(dx)$ if $a \in A$ because $|x|$ is bounded on $(-a, +a)$. Thus

$$\left|\int_{-\infty}^{+\infty} |x|(\mu_n(dx) - \mu_0(dx))\right| \\ \leq \int_{-a}^{a} |x|(\mu_n(dx) - \mu_0(dx)) + \int_{K_a^c} |x|\mu_n(dx) + \int_{K_a^c} |x|\mu_0(dx).$$

But, by Schwarz's inequality and part (1),

$$\left(\int_{K_a^c} |x|\mu_n(dx)\right)^2 \leq \left(\int_{K_a^c} x^2\mu_n(dx)\right)\int_{K_a^c} \mu_n(dx) \leq M\mu(K_n^2) \leq \frac{M^2}{a^2}.$$

Hence, if $a \in A$,

$$\limsup_{n\to\infty} \left| \int_{-\infty}^{+\infty} |x|(\mu_n(dx) - \mu(dx)) \right| \le \frac{2M}{a}.$$

Since A is unbounded, this proves (2).
(3) If δ_a is the unit mass at a, we define

$$\begin{aligned} \mu_0 &= \delta_0 \\ \mu_n &= (1 - \tfrac{1}{n})\delta_0 + \tfrac{1}{n}\delta_{\sqrt{n}} \quad \text{for } n > 0. \end{aligned}$$

Then μ_n converges weakly to μ_0 and $\displaystyle\int_{-\infty}^{+\infty} x^2 \mu_n(dx) = 1$ if $n > 0$, but $\displaystyle\int_{-\infty}^{+\infty} x^2 \mu_0(dx) = 0$.

Problem II-15. *If g is a measurable function on $(0, +\infty)$ which is locally integrable, and if $A = \lim_{T\to+\infty} \int_1^T g(x)dx$ and $B = \lim_{\epsilon\to 0} \int_\epsilon^1 g(x)dx$ exist, we say that $\int_0^{+\infty} g(x)dx$ exists and equals $A + B$.*

Let f be measurable and locally integrable on $(0, +\infty)$ and suppose that $\lim_{T\to+\infty} \int_1^T f(x)\frac{dx}{x}$ exists. Let a and b be positive.
(1) Suppose that $K = \int_0^\infty f(x)\frac{dx}{x}$ exists and let F be defined by $F(x) = \int_1^x f(t)dt$. Show that $\int_0^\infty [F(ax) - F(bx)]\frac{dx}{x^2}$ exists and express the integral in terms of a, b, and K.
(2) Suppose that $L = \lim_{\epsilon\to 0} f(x)$ exists. Show that $\int_0^\infty (f(ax) - f(bx))\frac{dx}{x}$ exists and express the integral in terms of a, b, and L.

SOLUTION.
(1)

$$\begin{aligned} \int_1^T (F(ax) - F(bx))\frac{dx}{x^2} &= \int_1^T \frac{dx}{x^2}\int_{bx}^{ax} f(t)dt &&= \int_1^T \frac{dx}{x}\int_a^b f(xu)du \\ &= \int_a^b \int_1^T f(xu)\frac{dx}{x} &&= \int_a^b du \int_u^{Tu} f(v)\frac{dv}{v}, \end{aligned}$$

with obvious changes of variable and the use of Fubini in the third equality. Let $G(x) = \int_1^x F(v)\frac{dv}{v}$; then G is continuous on $(0, +\infty)$ and $\lim_{x\to+\infty} G(x)$ exists. Hence $\sup_{T\ge 1} \int_u^{Tu} f(v)\frac{dv}{v} = \sup_{T\ge 1}[G(Tu) - G(u)]$ is a bounded function of u on $[a, b]$, and it follows from dominated convergence that

$$\lim_{T\to+\infty} \int_1^T (F(ax) - F(bx))\frac{dx}{x^2} = \int_a^b du \int_u^\infty f(v)\frac{dv}{v}.$$

Similarly, $\lim_{\epsilon\to 0} \int_\epsilon^1 (F(ax) - F(bx))\frac{dx}{x^2} = \int_a^b du \int_0^u f(v)\frac{dv}{v}$, and hence

$$\int_0^{+\infty} (F(ax) - F(bx))\frac{dx}{x^2} = K(b - a).$$

(2) Since $\int_1^\infty f(ax)\frac{dx}{x}$ and $\int_1^\infty f(bx)\frac{dx}{x}$ exist, obvious changes of variable give $\int_1^\infty [f(ax) - f(bx)]\frac{dx}{x} = \int_a^b f(u)\frac{du}{u}$. Hence

$$\int_\epsilon^1 [f(ax)-f(bx)]\frac{dx}{x} = \int_{a\epsilon}^a f(u)\frac{du}{u} - \int_{b\epsilon}^b f(u)\frac{du}{u} = \int_{a\epsilon}^{b\epsilon} f(u)\frac{du}{u} - \int_a^b f(u)\frac{du}{u}.$$

However, $\int_{a\epsilon}^{b\epsilon} f(u)\frac{du}{u} = \int_a^b f(\epsilon v)\frac{dv}{v}$. Since, for every $\delta > 0$, there exists $k_\delta > 0$ such that $0 < x \leq k_\delta$ implies $|f(x) - L| \leq \delta$, we conclude that $\sup_{\epsilon \leq k_1} |f(\epsilon v)|$ is bounded. By dominated convergence,

$$\lim_{\epsilon \to 0} \int_a^b f(\epsilon v)\frac{dv}{v} = L \int_a^b \frac{dv}{v} = L \log\frac{b}{a} = \int_0^\infty (f(ax) - f(bx))\frac{dx}{x}.$$

Problem II-16. *Writing $x^{-1} = \int_0^\infty e^{-yx}dy$ for $x > 0$ and applying Fubini's theorem, show that the integral $\int_0^\infty \sin x \frac{dx}{x}$ exists (in the sense of Problem II-15) and compute it. Use this to evaluate the integrals $\int_0^\infty (\cos ax - \cos bx)\frac{dx}{x^2}$ and $\int_0^\infty (\cos ax - \cos bx)\frac{dx}{x}$ if a, $b > 0$. (See Problem II-15.)*

Solution. If $T > 0$,

$$\begin{aligned} F(T) &= \int_0^T \sin x \frac{dx}{x} = \int_0^T \sin x dx \int_0^\infty \mathrm{e}^{-xy}dy \\ &= \int_0^\infty dy \int \mathrm{e}^{-xy} \sin x dx \\ &= \int_0^\infty \frac{1}{y^2+1}(1 - \mathrm{e}^{-yT}(\cos T + y\sin T))dy. \end{aligned}$$

But $\sup_{T \geq 1} |\mathrm{e}^{-yT}(\cos T + y\sin T)| \leq \mathrm{e}^{-y}(1+y)$. Hence, by dominated convergence, $\lim_{T \to +\infty} F(T) = \int_0^\infty \frac{dy}{1+y^2} = \frac{\pi}{2}$. Applying Problem II-15(1) to $f(x) = \sin x$, we obtain

$$\int_0^\infty (\cos ax - \cos bx)\frac{dx}{x^2} = (b-a)\frac{\pi}{2}.$$

We next apply Problem II-15(2) to $f(x) = \cos x$; it is applicable since $\int_{\frac{\pi}{2}}^\infty \frac{\cos x}{x}dx$ exists. Setting $x = \frac{\pi}{2} + t$ gives

$$\int_{\frac{\pi}{2}}^\infty \frac{\cos x}{x}dx = \int_0^\infty \frac{\sin t}{t}dt - \frac{\pi}{2}\int_0^\infty \frac{\sin t}{t(t+\frac{\pi}{2})}dt.$$

Hence $\int_0^\infty (\cos ax - \cos bx)\frac{dx}{x} = \log\frac{b}{a}$.

Problem II-17. *For an interval I in $\mathbf{R}$, $L^p(I)$ denotes the set of real-valued functions (rather, equivalence classes of functions) whose p^{th} power is integrable with respect to Lebesgue measure on I.*
(1) Show that $L^{p'}([0,1]) \subset L^p([0,1])$ if $0 < p < p' \leq \infty$. Give an example of a function in $L^1([0,1]) \setminus L^2([0,1])$.
(2) Give examples of functions in $L^1(\mathbf{R}) \setminus L^2(\mathbf{R})$ and in $L^2(\mathbf{R}) \setminus L^1(\mathbf{R})$.
(3) ℓ^p is the set of real-valued sequences $a = \{a_n\}_{n\geq 0}$ such that $\sum |a_n|^p < \infty$. Show that $\ell^{p'}(\mathbf{N}) \supset \ell^p(\mathbf{N})$ if $0 < p < p' \leq \infty$. Give an example of a sequence in $\ell^2 \setminus \ell^1(\mathbf{N})$.

SOLUTION.
(1) If $f \in L^{p'}([0,1])$, we apply Jensen's inequality I-9.2.2 to the convex function $\varphi(y) = |y|^{\frac{p'}{p}}$ and to $g = |f|^p$:

$$\varphi\left[\int_0^1 g(x)dx\right] \leq \int_0^1 \varphi[g(x)]dx < \infty.$$

If $f(x) = |x|^{-\frac{1}{2}}$, then $f \in L^1([0,1]) \setminus L^2([0,1])$.
(2) If $f(x) = |x|^{-\frac{1}{2}}(1+|x|)^{-1}$, then $f \in L^1(\mathbf{R}) \setminus L^2(\mathbf{R})$.
If $f(x) = (1+|x|)^{-1}$, then $f \in L^2(\mathbf{R}) \setminus L^1(\mathbf{R})$.
(3) If $a \in \ell^p$ with $p < \infty$, then $a_n \to 0$ as $n \to \infty$; hence $\sup_n |a_n| = m$ is finite. Thus $\sum_n |a_n|^{p'} \leq m \sum_n |a_n| < \infty$.
If $a_n = \frac{1}{n+1}$, then $a \in \ell^2(\mathbf{N}) \setminus \ell^1(\mathbf{N})$.

Problem II-18. *Let $\mathbf{R}_+^{n+1}$ denote the set of pairs (a,p) with $p > 0$ and $a \in \mathbf{R}^n$. Euclidean space $\mathbf{R}^n$ is equipped with the scalar product $\langle a, t\rangle$ and the norm $\|a\|$. Let*

$$K(a,p) = K_n p\left[\|a\|^2 + p^2\right]^{-\frac{n+1}{2}},$$

where K_n is the constant such that $\int_{\mathbf{R}^n} K(x,1)dx = 1$. The goal of this problem is to calculate

$$I_t(a,p) = \int_{\mathbf{R}^n} \exp i\langle x,t\rangle K(x-a,p)dx,$$

where $t \in \mathbf{R}^n$.

If $f : \mathbf{R}_+^{n+1} \to \mathbf{C}$, we write $D_0 f = \frac{\partial}{\partial p} f$ and $D_j f = \frac{\partial}{\partial a_j} f$ for $j = 1, \ldots, n$. f is said to be harmonic *in $\mathbf{R}_+^{n+1}$ if*

$$(D_0^2 + \cdots + D_n^2) f(a,p) = 0 \quad \textit{for every } (a,p) \in \mathbf{R}_+^{n+1}.$$

(1) Show that K is harmonic in $\mathbf{R}_+^{n+1}$. Show that, if $p_0 > 0$ and $V = (\frac{p_0}{2}, \frac{3p_0}{2})$, there exists a constant C such that $|D_i K(a,p)|$ and $|D_i D_j K(a,p)|$ are less than $C(1+\|a\|^2)^{-\frac{n+1}{2}}$ for all $(a,p) \in \mathbf{R}^n \times V$ and $i, j = 0, 1, \ldots, n$.

(2) Let μ be a Radon measure on $\mathbf{R}^n$ such that

$$\int_{\mathbf{R}^n} (1+\|x\|^2)^{-\frac{n+1}{2}} |\mu|(dx) < \infty$$

and let $F_\mu(a,p) = \int_{\mathbf{R}^n} K(x-a,p)\mu(dx)$. Show that F_μ is harmonic and that $\lim_{p\to+\infty} F_\mu(a,p) = 0$.
(3) Show that there exists a function $g : \mathbf{R}^n \to \mathbf{C}$ such that $I_t(a,p) = g(pt)\exp(i\langle a,t\rangle)$.
Use (2) to calculate g.

SOLUTION. (1) For simplicity, we set $f(\|a\|,p) = K(a,p)$; the function $f(y,p)$ is thus defined for $y \geq 0$ and $p > 0$. Then

$$(D_1^2 + \cdots + D_n^2)K(a,p) = \frac{\partial^2}{\partial y^2} f(\|a\|,p) + \frac{n-1}{\|a\|}\frac{\partial}{\partial y} f(\|a\|,p).$$

A calculation gives

$$\frac{\partial}{\partial y} f = -\frac{(n+1)y}{p^2+y^2} f, \quad \frac{\partial^2}{\partial y^2} f = \frac{(n+1)(n+2)y^2 - (n+1)p^2}{(p^2+y^2)^2} f,$$

and hence

$$(i) \qquad \frac{\partial^2}{\partial y^2} f + \frac{n-1}{y}\frac{\partial}{\partial y} f = \frac{3(n+1)y^2 - (n+1)np^2}{(p^2+y^2)^2} f.$$

Similarly, $\frac{\partial}{\partial p} f = \frac{y^2 - np^2}{p(p^2+y^2)} f$, and

$$(ii) \qquad \frac{\partial^2}{\partial p^2} f = -\frac{3(n+1)y^2 - (n+1)np^2}{(p^2+y^2)^2} f.$$

(i) and (ii) show that K is harmonic.

Next, it is easy to check that the functions $\frac{1}{K} D_i K$ and $\frac{1}{K} D_i D_j K$ are bounded on $\mathbf{R}^n \times V$. Since $\sup_{y \geq 0} f(p,y)(1+y^2)^{-\frac{n+1}{2}}$ is a continuous function of p, it is bounded on V. Hence there exists a constant C such that, for all i and j, $|(D_i K)(a,p)|$ and $|(D_i D_j)K(a,p)|$ are bounded above by $C(1+\|a\|^2)^{-\frac{n+1}{2}}$ in $\mathbf{R}^n \times V$.
(2) By (1), if V_1 is a compact neighborhood of (a_0,p_0) contained in $\mathbf{R}^n \times [\frac{p_0}{2}, \frac{3p_0}{2}]$, the following estimate holds for $(a,p) \in V_1$:

$$|(D_j K)(x-a,p)| \leq C(1+\|x-a\|^2)^{-\frac{n+1}{2}} \leq C_1(1+\|x\|^2)^{-\frac{n+1}{2}},$$

where C_1 is a constant. The conditions of the theorem on differentiating under the integral sign (I-7.8.4) are satisfied and we have

$$(D_j F_\mu)(a,p) = \int_{\mathbf{R}^n} (D_j K)(x-a,p)\mu(dx),$$

and similarly for the second derivatives. Hence

$$(D_0^2 + \cdots + D_n^2)(F_\mu)(a,p) = \int_{\mathbf{R}^n} (D_0^2 + \cdots + D_n^2)K(x-a,p)\mu(dx) = 0.$$

Finally, if $p > 1$,

$$|F_\mu(a,p)| \leq \frac{1}{p^n} \int_{\mathbf{R}} K(x-a,1)|\mu|(dx) \to 0 \quad \text{as } p \to \infty.$$

(3) Making the change of variable $x = a + pu$ in $I_t(a,p)$, we obtain

$$I_t(a,p) = \exp(i\langle a,t\rangle) \int_{\mathbf{R}^n} \exp(i\langle u, pt\rangle)K(u,1)du,$$

and hence $g(t) = I_t(0,1)$. Applying (2) to $\mu(dx) = \exp(i\langle x,t\rangle)dx$ shows that I_t is harmonic and that $I_t(a,p) \to 0$ as $p \to \infty$. Hence, with $g_t(p) = g(pt)$, harmonicity implies that

$$g_t''(p) - \|t\|^2 g_t(p) = 0 \quad \text{for } p > 0.$$

Thus $g_t(p) = A_t \exp(-p\|t\|) + B_t \exp(p\|t\|)$. Since $g_t(p) \to 0$ as $p \to \infty$, this shows that $B_t = 0$. Since $g_t(p)\exp(p\|t\|) = A_t$ is a function of pt, A_t is a constant; $A_t = 1$ since $I_0(0,1) = 1$; hence $I_t(a,p) = \exp(-p\|t\| + i\langle a,t\rangle)$.

REMARKS. 1. In n dimensions, $K(x-a,p)$ is sometimes called the Poisson kernel; in $\mathbf{R}^n$, it is sometimes called the Cauchy distribution.
2. The calculation giving $K_n = \Gamma(\frac{n+1}{2})\pi^{-\frac{n+1}{2}}$ is carried out in Problem III-4.
3. Another way of calculating $I_t(a,p)$ is given in Problem IV-13.
4. Another way of showing that $K(a,p)$ is harmonic is to introduce

$$\begin{aligned} G(a,p) &= \left[a^2 + p^2\right]^{-\frac{n-1}{2}} && \text{if } n > 1 \\ &= \log(a^2 + p^2) && \text{if } n = 1, \end{aligned}$$

to prove that G is harmonic on $\mathbf{R}_+^{n+1}$, and to observe that K is proportional to $\frac{\partial G}{\partial p}$.
5. We could use an analogous method to prove the identity

$$h(a,p) = \int_{\mathbf{R}^n} \exp(-p\|t\| + i\langle a,t\rangle)dt = Kp(\|a\|^2 + p^2)^{-\frac{n+1}{2}},$$

where $K = h(0,1)$, as follows. We begin by noting that, since $(a,p) \mapsto \exp(-p\|t\| + i\langle a,t\rangle)$ is harmonic in $\mathbf{R}_+^{n+1}$, so is h. We next observe, by a change of variable, that there must exist a function $g : [0,+\infty) \to \mathbf{R}$ such that $h(a,p) = p^{-n}g(\frac{\|a\|}{p})$; the harmonicity of h then leads to the following differential equation on $(0,+\infty)$ for g:

$$(1+y^2)g''(y) + ((2n+2)y + (n-1)y^{-1})g'(y) + n(n+1)g(y) = 0.$$

The function $g_1(y) = K(1+y^2)^{-\frac{n+1}{2}}$ is a solution of this equation. Taking $G = \frac{g}{g_1}$ as a new unknown function in the equation, we find that there exist constants A and B such that

$$G = A + B\int_y^1 (1+u^2)^{\frac{n-1}{2}} \frac{du}{u^{n-1}}.$$

Since $g(y) \to K$ as $y \to 0$ by definition, it follows that $A = 1$ and $B = 0$ if $n > 1$. The case $n = 1$ is treated a bit differently, by first finding $h(a,p) = \frac{Kp+B|a|}{p^2+a^2}$, then using the fact that $\frac{\partial h}{\partial a}(0,p)$ exists in order to see that $B = 0$.

Problem II-19. *(1) Let μ and ν be positive measures on $\mathbf{R}$ such that there exists an interval $[a,b] \subset \mathbf{R}$ with $\mu([a,b]) = \mu(\mathbf{R})$ and $\nu([a,b]) = \nu(\mathbf{R})$. Show that $\mu = \nu$ if and only if*

$$\int_{\mathbf{R}} x^n \mu(dx) = \int_{\mathbf{R}} x^n \nu(dx), \quad \forall n = 0, 1, 2, \ldots.$$

(2) Let μ be a positive measure on $[0,+\infty)$ (not necessarily bounded). Its Laplace transform *is the function from $\mathbf{R}$ to $[0,+\infty]$ defined by*

$$s \mapsto (L\mu)(s) = \int_0^\infty e^{-sx}\mu(dx).$$

(a) If $E_\mu = \{s : (L\mu)(s) < \infty\}$, show that E_μ is an interval which, if nonempty, is unbounded on the right. Give examples where $E_\mu = \mathbf{R}$, $\emptyset$, $(0,+\infty)$, and $[0,+\infty)$.
(b) Use (1) to show that if there exists a number a such that $L\mu = L\nu < \infty$ on $[a,+\infty)$, then $\mu = \nu$.

Solution. (1) The hypothesis implies that $\int P(x)[\mu - \nu](dx) = 0$ for every polynomial P. By the Stone-Weierstrass theorem, for every continuous function f on $[a,b]$ and for every $\epsilon > 0$ there exists a polynomial P_ϵ such that $\sup_{x\in[a,b]} |f(x) - P_\epsilon(x)| \leq \epsilon$. Hence $\int f\mu(dx) = \int f\nu(dx)$, and by Riesz's theorem $\mu = \nu$. (See II-2.2.1.)
(2a) If $s_0 \in E_\mu$ and $s > s_0$, it is clear that $s \in E$ since the function $\mathrm{e}^{-(s-s_0)x}$ is bounded on $[0,+\infty]$. As examples, we can take

$$\mu(dx) = \mathrm{e}^{\frac{x^2}{2}}dx, \quad \mathrm{e}^{-\frac{x^2}{2}}dx, \quad dx, \quad \text{and} \quad \mathrm{e}^{-\sqrt{x}}dx.$$

(b) By hypothesis, $\mu_1(dx) = \mathrm{e}^{-ax}\mu(dx)$ and $\nu_1(dx) = \mathrm{e}^{-ax}\nu(dx)$ are bounded measures. Their images $\mu_2(dx)$ and $\nu_2(dx)$ under the mapping $x \mapsto \mathrm{e}^{-x}$ are concentrated in $[0,1]$. Since

$$(L\mu)(a+n) = \int_0^\infty e^{-xn-ax}\mu(dx) = \int_0^1 x^n \mu_2(dx)$$

if $n \in \mathbf{N}$, it follows from (1) that $\mu_2 = \nu_2$ and hence that $\mu = \nu$.

Problem II-20. *Give examples of sequences $\{\mu_n\}_{n=1}^\infty$ of positive Radon measures on $\mathbf{R}$ such that, for each sequence, there exists a positive Radon measure μ with $\lim_{n\to\infty} \mu_n = \mu$*

(1) vaguely but not weakly;

(2) weakly but not narrowly; and

(3) narrowly but not in norm.

SOLUTION. (1) $\mu_n(dx) = \mathbf{1}_{[-n,n]}(x)dx$ converges vaguely to Lebesgue measure μ on $\mathbf{R}$; it cannot converge weakly because μ is unbounded.
(2) If δ_a is the Dirac measure at a, then $\mu_n = \delta_n$ converges weakly to the measure 0 but cannot converge narrowly since $\|\mu_n\| = 1$ and $\|0\| = 0$.
(3) $\mu_n = \delta_{\frac{1}{n}}$ converges narrowly to $\mu = \delta_0$, but $\|\mu_n - \mu\| = 2$ does not tend to 0.

REMARK. If the sequence of positive measures $\{\mu_n\}_{n=1}^\infty$ converges vaguely to μ and $\mu(X) < \infty$, then $\mu_n \to \mu$ weakly, since $C_K(X)$ is dense in $C_0(X)$. It should also be noted that narrow and weak convergence coincide when X is complete.

Problem II-21. *Let X be a locally compact space that is countable at infinity and let $M^1(X)$ be the set of signed Radon measures ν on X such that $|\nu|$ has finite total mass $\|\nu\|$. If $\{\nu_n\}_{n=1}^\infty$ is a sequence in $M^1(X)$ such that $r = \sup_n \|\nu_n\| < \infty$, show that there exist ν in $M^1(X)$ and an increasing sequence of integers $\{n_k\}_{k=1}^\infty$ such that $\nu_{n_k} \to \nu$ as $k \to \infty$. Show also that $\nu \geq 0$ if $\nu_n \geq 0$ for every n.*

METHOD. Use Theorem II-6.6.

SOLUTION. Since $M^1(X)$ is the dual of $C_0(X)$, by the Banach-Alaoglu theorem the closed ball of radius r in $M^1(X)$ is compact in the weak topology; this implies the result.

Moreover, if $\nu_n \geq 0$ for every n, let $f \in C_0(X)$ with $f \geq 0$. Then

$$\lim_{n\to\infty} \int f\nu_n = \int f\nu \geq 0,$$

which shows that ν is positive.

REMARK. When $X = \mathbf{R}$, $\nu_n \geq 0$, and $r = 1$, this property is often called Helly's theorem.

Problem II-22. *On a locally compact space X that is countable at infinity, let μ and $\{\mu_n\}_{n=1}^{\infty}$ be positive Radon measures such that μ_n converges vaguely to μ as $n \to \infty$.*
(1) If O is an arbitrary open set, show that $\mu(O) \leq \liminf_{n\to\infty} \mu_n(O)$.
(2) Suppose that O is an open set with compact closure K and such that its boundary $\partial O = K \setminus O$ has μ-measure 0. Let $\{O_k\}_{k=1}^{\infty}$ be a decreasing sequence of open subsets of X such that $\cap_{k=1}^{\infty} O_k = K$ and let f_k be a function equal to 1 on K and to 0 on O_k^c, and such that $0 \leq f(x) \leq 1$ for x in O_k. (Such a function exists by Urysohn's lemma, II-1.1.) Show that

$$\limsup_{n\to\infty} \mu_n(O) \leq \int_X f_k(x)\mu(dx),$$

and conclude that $\mu_n(O) \to \mu(O)$ as $n \to \infty$.
(3) If μ and $\{\mu_n\}_{n=1}^{\infty}$ are Radon measures on $\mathbf{R}$, positive and with total mass less than or equal to 1, show that μ_n converges weakly to μ as $n \to \infty$ if and only if

$$\mu_n((a,b)) \to \mu((a,b)) \quad \textit{as } n \to \infty$$

for all points of continuity of the distribution function $x \mapsto \mu((-\infty, x))$.

If moreover $\mu_n(\mathbf{R}) = \mu(\mathbf{R}) = 1$, show that $\mu_n \to \mu$ narrowly if and only if

$$\mu_n((-\infty,x)) \to \mu((-\infty,x)) \quad \textit{as } n \to \infty$$

for every point of continuity of the right-hand side.

METHOD. Use Problems II-1 and II-21 together with Theorem II-6.8.

SOLUTION. As in II-2.3.2, we write

$$T(O) = \{f \in C_K(X) : \operatorname{supp}(f) \subset O, \quad 0 \leq f(x) \leq 1 \quad \forall x \in O\}.$$

(1) For every g in $T(O)$,

$$\int g\mu_n \leq \sup\left\{\int f\mu_n : f \in T(O)\right\} = \mu_n(O)$$

by the definition of $\mu_n(O)$. (See II-2.4.1.) Hence

$$\int g\mu = \lim_{n\to\infty} \int g\mu_n \leq \liminf_{n\to\infty} \mu_n(O) \quad \text{and}$$

$$\mu(O) = \sup\left\{\int g\mu : g \in T(O)\right\} \leq \liminf \mu_n(O).$$

(2) $\mu_n(O) \leq \int f_k\mu_n$ and hence

$$\limsup_{n\to\infty} \mu_n(O) \leq \lim_{n\to\infty} \int f_k\mu_n = \int f_k\mu.$$

However, $\mu(O_k) \to \mu(K)$ as $k \to \infty$ by the definition of $\mu(K)$. (See II-2.4.2.) Since $\mu(K \setminus O) = 0$ and $\int f_k \mu \leq \mu(O_k)$,

$$\limsup_{n\to\infty} \mu_n(O) \leq \mu(O).$$

Comparing this with (1), we have indeed shown that $\lim_{n\to\infty} \mu_n(O) = \mu(O)$.

(3) Let C be the set of points of continuity of the mapping $x \mapsto \mu((-\infty, x))$. If $\mu_n \to \mu$ weakly and if a, $b \in C$, then $\mu(\{a\}) = \mu(\{b\}) = 0$. Applying (2) to the open set $O = (a, b)$ shows that $\lim_{n\to\infty} \mu_n(O) = \mu(O)$. Conversely, let ν be a positive measure with total mass ≤ 1 and let $\{n_k\}_{k=1}^{\infty}$ be an increasing sequence of integers such that μ_{n_k} converges weakly to ν as $k \to \infty$. (The existence of such a sequence is guaranteed by Problem II-21.) Then, by the forward implication, if C_1 is the set of points of continuity of $x \mapsto \nu((-\infty, x))$, we have $\mu_n((a, b)) = \nu((a, b))$ whenever a, $b \in C_1 \cap C$. Since $\mathbf{R} \setminus C_1 \cap C$ is countable, it follows from the last inequality and Problem II-1 that $\mu = \nu$. Since every weakly convergent subsequence of $\{\mu_n\}_{n=1}^{\infty}$ converges to μ, we conclude that μ_n converges weakly to μ.

Next, it is clear that if $\mu_n((-\infty, x)) \to \mu((-\infty, x))$ as $n \to \infty$ for every x in C, then

$$\mu_n((a, b)) \to \mu((a, b)) \quad \text{as } n \to \infty$$

for all a, $b \in C$. Hence μ_n converges weakly to μ as $n \to \infty$. If moreover $\mu_n(\mathbf{R}) = \mu(\mathbf{R}) = 1$, then the convergence is narrow by Theorem II-6.8 ((v) $\Rightarrow$ (ii)). The converse is a bit more delicate. By the same theorem ((ii) $\Rightarrow$ (iv)), for every $\epsilon > 0$ there exist a compact interval $[a_\epsilon, b_\epsilon]$ and an integer $N(\epsilon)$ such that $\mu_n([a_\epsilon, b_\epsilon]) \geq 1 - \epsilon$. We may take a_ϵ in $C \cap \bigcap_{n=1}^{\infty} C_n$, where C_n is the set of points of continuity of $x \mapsto \mu_n((-\infty, x))$; hence $\mu_n((a_\epsilon, +\infty)) \geq 1 - \epsilon$ if $n \geq N(\epsilon)$. It follows that if $[x \in C]$ and $n \geq N(\epsilon)$, then

$$\begin{aligned} |\mu_n((-\infty, x)) - \mu((-\infty, x))| &\leq 2\epsilon & \text{if} \quad x \leq a_\epsilon \\ |\mu_n((-\infty, x)) - \mu((-\infty, x))| &\leq 2\epsilon + |\mu_n((a_\epsilon, x) - \mu((a_\epsilon, x))| & \text{if} \quad x > a_\epsilon \end{aligned}$$

The last quantity tends to 2ϵ as $n \to +\infty$. Hence $\lim_{n\to\infty} \mu_n((-\infty, x)) = \mu((-\infty, x))$ for every x in C.

REMARK. In practice, (3) gives a necessary and sufficient condition for the convergence of probability distributions on $\mathbf{R}$; it is often taken as a definition in elementary texts.

Problem II-23. *Let X be a locally compact space that is countable at infinity, and let μ and $\{\mu_n\}_{n=1}^{\infty}$ be Radon measures on X such that μ_n converges vaguely to μ.*

(1) If O is an open set in X and μ^* is the restriction of μ to O, show that μ_n^* converges vaguely to μ^* as $n \to \infty$.

(2) Show by an example that the statement is false if O is replaced by a closed set.
(3) Suppose that $X = \mathbf{R}$ and that $\mu_n \geq 0$, $n = 1, 2, \ldots$. Let a and b be real numbers with $a < b$. Show that there exist numbers p and q and an increasing sequence of integers $\{n_k\}_{k=1}^{\infty}$ such that, for every continuous function f on $[a, b]$,

$$\int_{[a,b]} f\mu_{n_k} \to pf(a) + qf(b) + \int_{[a,b]} f\mu \quad \textit{as } n \to \infty.$$

METHOD. Use Problem II-21.

SOLUTION. (1) Let $f : O \to \mathbf{R}$ be continuous with compact support, and let $\tilde{f}$ be defined by $\tilde{f}(x) = f(x)$ if $x \in O$ and $\tilde{f}(x) = 0$ otherwise. Then

$$\int_O f\mu_n^* = \int_X \tilde{f}\mu_n \to \int_X \tilde{f}\mu = \int_O f\mu^* \quad \text{as } n \to \infty.$$

(2) Take $X = \mathbf{R}$, $F = [0, 1]$, $\mu_n = \delta_{-\frac{1}{n}}$, and $\mu = \delta_0$. Then $0 = \mu^*$ does not converge vaguely to $\mu^* = \delta_0$.
(3) If h is continuous with compact support, with values in $[0, 1]$, and equal to 1 on $[a, b]$, then

$$\mu_n([a, b]) \leq \int h\mu_n \to \int h\mu \quad \text{as } n \to \infty.$$

Since $\{\mu_n([a, b])\}_{n=1}^{\infty}$ is a bounded sequence, $\{\mu_n^*\}_{n=1}^{\infty}$ converges to μ^* not just vaguely (see (1)) but weakly. (See the remarks on Problem II-20.) However, this convergence is not necessarily narrow. By Problem II-21, there exist a Radon measure ν on the compact set $[a, b]$ and an increasing sequence of integers $\{n_k\}_{k=1}^{\infty}$ such that the restriction of μ_{n_k} to $[a, b]$ converges narrowly to ν and hence $\mu_{n_k} \to \nu^*$ weakly as $k \to \infty$ (see (1)). Hence $\mu^* = \nu^*$, and it follows that, as $k \to \infty$,

$$\int_{[a,b]} f\mu_{n_k} \to f\nu = pf(a) + qf(b) + \int_{[a,b]} f\mu,$$

where $p = \nu(\{a\}) - \mu(\{a\})$ and $q = \nu(\{b\}) - \mu(\{b\})$.

REMARK. It is necessary to work with a subsequence in (3). For example, if $[a, b] = [0, 1]$ and $\mu_n = \delta_{(-1)^n n^{-1}}$,

$$\int_{[0,1]} f\mu_{2k} \to f(0) \quad \text{as } k \to \infty$$

although

$$\int_{[0,1]} f\mu_{2k+1} = 0 \quad \text{for every } k.$$

Problem II-24. *(1) Let O and O' be two open sets in $\mathbf{R}^n$, let f be a diffeomorphism from O onto O', and let φ be a measurable function on O' such that $\int_{O'} \varphi(x')dx' < \infty$. Show that*

$$\int_O \varphi(f(x))|detJ_f(x)|dx = \int_{O'} \varphi(x')dx',$$

where $|detJ_f(x)|$ is the Jacobian.
(2) Let $a \in \mathbf{R} \cup \{-\infty\}$. Let f and g be functions satisfying the following conditions: (i) f is continuously differentiable for $x > a$; (ii) g is defined and integrable on $[0, +\infty)$; (iii) $|f'(x + \frac{u^2}{2})| \leq g(u)$ for all $x > a$; and (iv) both $u \mapsto ug(u)$ and $u \mapsto f(x + \frac{u^2}{2})$ are integrable on $[0, +\infty)$. If $F(x) = \int_{-\infty}^{+\infty} f(x+\frac{u^2}{2})du$, show by a change of variables in polar coordinates that

$$f(x) = -\frac{1}{2\pi} \int_{-\infty}^{+\infty} F'(x + \frac{v^2}{2})dv.$$

SOLUTION. (1) If $\varphi \in C_K(O')$, this is Theorem II-4.4.1. Hence, by the Radon-Riesz theorem, $\mathbf{1}_{O'}(x')dx'$ is the direct image under f (in the sense of Problem I-14) of the measure $\mathbf{1}_O(x)|\det J_f(x)|dx$. The conclusion follows from Problem I-14.
(2) Since g is integrable, we can differentiate under the integral sign and write

$$-\frac{1}{2\pi} \int_{-\infty}^{+\infty} F'(x + \frac{v^2}{2})dv = -\frac{1}{2} \int_{-\infty}^{+\infty} dv \int_{-\infty}^{+\infty} f'(x + \frac{u^2+v^2}{2})dv.$$

We now apply (1) to $O = \mathbf{R}^2 \setminus [0, +\infty) \times \{0\}$, $O' = (0, 2\pi) \times (0, +\infty)$, and $f(u, v) = (\theta, \rho)$, with $u = \rho \cos\theta$, $v = \rho \sin\theta$, and $(\theta, \rho) = f'(x + \frac{\rho^2}{2})\rho$. The Jacobian is $|\det J_f(u, v)| = (u^2 + v^2)^{-\frac{1}{2}}$. The integral above can be written

$$-\frac{1}{2\pi} \int_0^{2\pi} d\theta \int_0^{\infty} f(x + \frac{\rho^2}{2})\rho d\rho = -\lim_{T\to\infty} \int_0^T f'(x + y)dy.$$

Since f' is integrable on $[0, +\infty)$, $\lim_{T\to\infty} f(T + x)$ exists. Since f is integrable, this limit is zero, and the result follows.

REMARK. The case $f(x) = e^{-x}$ is well known and is used in IV-4.3.2(i).

Problem II-25. *Consider a subset X of $\mathbf{R}^n$ with positive measure, a measurable function $f : X \to \mathbf{R}^n$, and a nonnegative locally integrable function*

h on X. Let μ denote the image in $\mathbf{R}^n$ of the measure $h(x)dx$ on X under f (in the sense of Problem I-14) if this image measure exists.
(1) If X and U are open sets and f is a diffeomorphism from X to U, show that

$$\mu(du) = h(f^{-1}(u))|detJ_{f^{-1}}(u)|du.$$

(2) If there exist an open subset U of $\mathbf{R}^n$ and disjoint open sets X_1, $X_2, \ldots, X_d$ contained in X such that the restriction f_j of f to X_j is a diffeomorphism on U, and if $X \setminus \sum_{j=1}^d X_j$ has Lebesgue measure zero, show that

$$\mu(du) = \sum_{j=1}^{d} h(f_j^{-1}(u))|detJ_{f_j^{-1}}(u)|\mathbf{1}_U(u)du.$$

(3) If $X = (0, +\infty)^2$, $c(x) = x^{-\frac{3}{2}}\exp[-(ax + \frac{b}{x})]$, $h(x,y) = c(x)c(y)$, and $f(x,y) = (u,v)$, with $u = x+y$ and $v = \frac{1}{x} + \frac{1}{y}$, calculate μ. Conclude from the result that the image of $hdxdy$ under the map $(x,y) \mapsto (x+y, \frac{1}{x} + \frac{1}{y} - \frac{4}{x+y})$ is also a product measure.

SOLUTION. (1) By definition, if φ is continuous with compact support in U,

$$\int_U \varphi(u)\mu(du) + \int_X \varphi(f(x))h(x)dx = \int_U \varphi(u)h(f^{-1}(u))|\mathrm{det}J_{f^{-1}}(u)|$$

by Problem II-24.
(2) If φ is continuous with compact support in U,

$$\begin{aligned}\int_U \varphi(u)\mu(du) &= \int_X \varphi(f(x))h(x)dx = \sum_{j=1}^{d}\int_{X_j} \varphi(f_j(x))h(x)dx \\ &= \sum_{j=1}^{d}\int_U \varphi(u)h(f_j^{-1}(u))|\mathrm{det}J_{f_j^{-1}}(u)|du.\end{aligned}$$

(3) Since (x,y) and (y,x) have the same image under f, f is not a diffeomorphism. Let $U = \{(u,v) : uv - 4 > 0,\ u > 0,\ v > 0\}$, let $X_{-1} = \{(x,y) : 0 < x < y\}$, and let $X_1 = \{(x,y) : 0 < y < x\}$. Elementary manipulations show that, for $\epsilon = \pm 1$, the restriction f_ϵ of f to X_ϵ is a diffeomorphism. Thus, if

$$(x,y) = f_\epsilon^{-1}(u,v),$$

then $x+y = u$, $xy = \frac{u}{v}$, and x and y are the roots of the following equation in r:

$$r^2 - ur + \frac{v}{u} = 0.$$

Hence $x = \frac{1}{2}(u + \epsilon\sqrt{u^2 - \frac{4u}{v}})$ and $y = \frac{1}{2}(u - \epsilon\sqrt{u^2 - \frac{4u}{v}})$. It is possible to calculate $|\mathrm{det}J_{f_{\epsilon^{-1}}}(u,v)|$ from these last two formulas. However, it is simpler to observe that

$$|\det J_{f_\epsilon^{-1}}(u,v)|\ |\det J_{f_\epsilon}(f_\epsilon^{-1}(u,v))| = 1.$$

Since

$$J_{f_\epsilon}(x,y) = \begin{pmatrix} 1 & 1 \\ -x^{-2} & -y^{-2} \end{pmatrix},$$

$$|\det J_{f_\epsilon}(x,y)| = \frac{x+y}{x^2y^2}|y-x| = \frac{1}{v^{\frac{3}{2}}\sqrt{v-\frac{4}{u}}}$$

independently of ϵ, since $|y-x| = \sqrt{u^2 - \frac{4u}{v}}$. Similarly,

$$h\left[f_\epsilon^{-1}(u,v)\right] = \frac{v^{\frac{3}{2}}}{u^{\frac{3}{2}}}\exp(-(au+bv))$$

for $\epsilon = \pm 1$. Hence, by (2),

$$\mu(du,dv) = \frac{2}{u^{\frac{3}{2}}\sqrt{v-\frac{4}{u}}}\exp(-(au+bv))\mathbf{1}_U(u,v)dudv.$$

(Note the factor 2.)

It is easier to find the image ν of μ under the map $g : (u,v) \to (u, v-\frac{4}{u}) = (u_1, v_1)$ because g is a diffeomorphism of U onto $X = (0,\infty)^2$. Since

$$J_g(u,v) = \begin{pmatrix} 1 & \frac{4}{u^2} \\ 0 & 1 \end{pmatrix},$$

$|\det J_g(u,v)| = 1$ and hence $|\det J_{g^{-1}}(u_1,v_1)| = 1$. It follows that

$$\begin{aligned} \nu(du_1,dv_1) &= \frac{2}{u_1^{\frac{3}{2}}}\exp(-(au_1 + \frac{4v}{u_1}))\mathbf{1}_{(0,+\infty)}(u_1)du_1 \\ &\times \frac{1}{v_1^{\frac{1}{2}}}\exp(-bv_1)\mathbf{1}_{(0,+\infty)}(v_1)dv_1. \end{aligned}$$

It is a remarkable fact that this is a product measure.

REMARKS. 1. The use of the change-of-variables theorem (II-4.4.1) to calculate the image of a measure is important in practice, especially in probability theory.

2. Problem II-12 treats a special case of (2) for $n = 1$.

3. (3) shows that if X and Y are independent random variables of density $Kc(x)dx$ (a distribution called "inverse Gaussian"), then $X+Y$ and $\frac{1}{X} + \frac{1}{Y} - \frac{4}{X+Y}$ are independent. It seems difficult to justify this result by Fourier analysis.

4. Calculating the image of $h(x)dx$ if f maps X into $\mathbf{R}^m$ with $m < n$ is more delicate. In practice, one applies (2) to $f_1(x) = (f(x), x_{m+1}, \ldots, x_n)$ and integrates with respect to $x_{m+1}, \ldots, x_n$ in order to find the image under f.

III

Fourier Analysis

Problem III-1. *Let G be the group O_d of $d \times d$ orthogonal matrices, acting on the Euclidean space $\mathbf{R}^d$. The scalar product and the norm are denoted by $\langle x, t\rangle$ and $\|t\|$, respectively. Let μ be a bounded complex measure on $\mathbf{R}^d$, with Fourier transform*

$$\widehat{\mu}(t) = \int_{\mathbf{R}^d} \exp(i\langle x, t\rangle)\mu(dx) \quad (t \in \mathbf{R}^d).$$

Prove the equivalence of the following three properties.
(1) μ is invariant under every element of G.
(2) There exists $\varphi : [0, \infty) \to \mathbf{C}$ such that $\widehat{\mu}(t) = \varphi(\|t\|)$ for every t.
(3) The image ν_a in $\mathbf{R}$ of μ under the mapping $x \mapsto \langle a, x\rangle$ does not depend on a when a ranges over the unit sphere S_{d-1} of $\mathbf{R}^d$.

SOLUTION.
(2) $\Rightarrow$ (1). If $g \in G$, its adjoint g^*, defined by $\langle gx, t\rangle = \langle x, g^*t\rangle$, is such that $g^* = g^{-1} \in G$. Hence

$$\widehat{\mu}(t) = \varphi(\|t\|) = \varphi(\|g^*t\|) = \widehat{\mu}(g^*t) = \int_{\mathbf{R}} \exp(i\langle gx, t\rangle)\mu(dx) = \widehat{\mu}_g(t)$$

is the image of μ under g. By the theorem on uniqueness of the Fourier transform (see the corollary to III-2.8), $\mu = \mu_g$.
(1) $\Rightarrow$ (2). $\widehat{\mu}_g(t) = \widehat{\mu}(g^*(t)) = \widehat{\mu}(t)$. If $u \geq 0$, let $t_u = (u, 0, \ldots, 0) \in \mathbf{R}^d$ and set $\varphi(u) = \widehat{\mu}(t_u)$. Since $\|t\| = u$ implies the existence of a g in G such that $g^*(t) = t_u$, it follows that $\widehat{\mu}(t) = \varphi(\|t\|)$.

(3) ⇒ (2). Let $\nu = \nu_a$ for every a in S_{d-1}. Then, if $t \neq 0$,

$$\widehat{\mu}(t) = \int_{\mathbf{R}^d} \exp(i\langle x, \frac{t}{\|t\|}\rangle \|t\|)\mu(dx) = \int_{\mathbf{R}} \exp(iy\|t\|)\nu(dy) = \widehat{\nu}(\|t\|).$$

(2) ⇒ (3). Since, for all nonzero t, $\widehat{\nu}_{\frac{t}{\|t\|}}(\|t\|) = \varphi(\|t\|)$, it follows that $\widehat{\nu}_a(s) = \varphi(s)$. But $\widehat{\nu}_{-a}(s) = \widehat{\nu}_a(-s)$ and $\widehat{\nu}_a$ and φ are continuous. Hence $\widehat{\nu}_a(-s) = \varphi(s)$ if $s > 0$ and $\widehat{\nu}_a(0) = \varphi(0)$. Since the Fourier transform of ν_a does not depend on a, neither does ν_a.

REMARK. Naturally, if μ is real, then $\widehat{\mu}(t) = \overline{\widehat{\mu}(-t)}$ implies that φ is real. But $\mu \geq 0$ does not imply that $\varphi \geq 0$. Thus, if σ is the uniform probability measure on S_2, the unit sphere in $\mathbf{R}^3$, $\widehat{\sigma}(t) = \frac{\sin \|t\|}{\|t\|}$.

Problem III-2. *Let T be a compact space, let G be a compact topological group, and let $(g,t) \mapsto gt$ be a continuous map from $G \times T$ to T such that $g \mapsto \{(g,t) \mapsto gt\}$ is a homomorphism from G to the group of bijections of T. Finally, suppose that (G,T) is a homogeneous space; that is, for every t_1 and t_2 in T there exists g such that $gt_1 = t_2$. Let dg denote the unique measure of total mass 1 on G which is invariant under left and right multiplication. (We accept without proof the existence and uniqueness of dg.)*
(1) If f is continuous on T, show that $t \mapsto \int_G f(g^{-1}t)dg$ is a constant $\sigma[f]$. Conclude that $\sigma[f]$ defines a probability measure on T which is invariant under the action of G.
(2) If μ is a probability measure on T which is invariant under the action of G, show that $g \mapsto \int_T f[g^{-1}t]\mu(dt)$ is a constant. Integrate with respect to dg and conclude that $\mu = \sigma$.
(3) If $(X, \mathcal{A})$ is an arbitrary measurable space and T is equipped with its Borel algebra, let $T \times X$ be given the product σ-algebra. Suppose that G acts on $T \times X$ by $g(t,x) = (gt,x)$. Show that every positive measure μ on $T \times X$ which is invariant under the action of G has the form $\sigma(dt) \otimes \nu(dx)$, where ν is a measure ≥ 0 on $(X, \mathcal{A})$. Converse?
METHOD. If $A \in \mathcal{A}$ is such that $\mu(T \times A) \in (0, +\infty)$, show that $\mu_A(B) = \frac{\mu(B \times A)}{\mu(T \times A)}$ defines a probability measure on T which is invariant under G.

(4) Apply the preceding results when $T = S_d$ is the unit sphere of the Euclidean space $\mathbf{R}^{d+1}$, where $G = O_{d+1}$ is the group of $(d+1) \times (d+1)$ orthogonal matrices and $X = (0, +\infty)$. Conclude that a probability measure P on $\mathbf{R}^{d+1} \setminus 0$ is invariant under G if and only if $\frac{x}{\|x\|}$ and $\|x\|$ are independent and $\frac{x}{\|x\|}$ has the uniform distribution on S_d.

SOLUTION. (1) σ defines a positive linear functional on the space of continuous functions on T and $\sigma(1) = 1$. By Riesz's theorem (II-2.2), it has a corresponding probability measure σ which is invariant under G.

(2) $\int_T f(g^{-1}t)\mu(dt) = \int_T f(t)\mu(dt)$ since μ is invariant under G. Using Fubini,

$$\int_T f(t)\mu(dt) = \int_T \mu(dt) \int_G f[g^{-1}t]dg = \sigma(f)\mu(T).$$

Since $\mu(T) = 1$, it follows that $\mu = \sigma$.
(3) It is clear that μ_A is invariant under G. Hence, by (2), $\mu_A = \sigma$. Set $\nu(A) = \mu(T \times A)$; then $\mu(B \times A) = \sigma(B)\nu(A)$. Extending the definition of ν to $\mathcal{A}$ by $\nu(A) = \mu(T \times A)\ \forall A$ gives $\mu = \sigma \otimes \nu$. The converse is obviously true.
(4) $\mathbf{R}^{d+1} \setminus \{0\}$ is homeomorphic to $S_d \times (0, +\infty)$ under the map $x \mapsto (\frac{x}{\|x\|}, \|x\|)$. All the rotation-invariant measures $\mu \geq 0$ are thus of the form $\sigma \otimes \nu$. If μ is a probability measure P, this is equivalent to the independence of $\frac{x}{\|x\|}$ and $\|x\|$ when $\mathbf{R}^{d+1} \setminus \{0\}$ is probabilized by P.

Problem III-3. *In the Euclidean space $\mathbf{R}^d$ equipped with the norm $\|x\|$, let m be Lebesgue measure.*
(1) If ν_0 and ν_1 are the images of m in $[0, +\infty)$ under the mappings $x \mapsto \|x\|$ and $x \mapsto \frac{\|x\|^2}{2}$ (see Problem I-14), show that

$$\nu_1(d\gamma) = \frac{(\sqrt{2\pi})^d}{\Gamma(\frac{d}{2})} \gamma^{\frac{d}{2}-1} d\gamma,$$

where Γ is the usual Euler function (see, for example, Problem IV-11). Use this to find $\nu_0(d\rho)$.
METHOD. Use the formula

$$\frac{1}{\sigma^d (\sqrt{2\pi})^d} \int_{\mathbf{R}^d} \exp\left(-\frac{\|x\|^2}{2\sigma^2}\right) dx = 1,$$

which holds for all $\sigma > 0$, to calculate the Laplace transform $(L\nu_1)(s)$ defined in Problem II-19.

(2) Keep the same notation m and ν_0 for the restrictions of m and ν_0 to $\mathbf{R}^d \setminus \{0\}$ and $(0, +\infty)$. If μ is a measure ≥ 0 on $\mathbf{R}^d \setminus \{0\}$ which has density f with respect to m, use Problem III-2 to show that the image of μ on $(0, +\infty)$ under the map $x \mapsto \|x\|$ is of the form $f_1(\rho)\nu_0(d\rho)$. Calculate the function f_1 in terms of f. If μ is rotation invariant, show that there exists a function $f_1 : (0, +\infty) \to [0, +\infty)$ such that $f_1(\|x\|) = f(x)$ m-a.e.

SOLUTION. (1) ν_0 and ν_1 exist even though m is unbounded, since the inverse images of compact sets are compact under the maps $x \mapsto \frac{\|x\|^2}{2}$ and $x \mapsto \|x\|$. By the formula above, for all $\sigma > 0$,

$$(\sqrt{2\pi})^d \sigma^d = \int_{\mathbf{R}^d} \exp\left(-\frac{\|x\|^2}{2\sigma^2}\right) dx = \int_0^\infty \exp\left(-\frac{\gamma}{\sigma^2}\right) \nu_1(d\gamma).$$

Hence, for all $s > 0$,

$$(L\nu_1)(s) = \int_0^\infty e^{-s\gamma}\nu_1(d\gamma) = (\sqrt{2\pi})^d s^{-\frac{d}{2}} = \frac{(\sqrt{2\pi})^d}{\Gamma(\frac{d}{2})}\int_0^{+\infty} e^{-s\gamma}\gamma^{\frac{d}{2}-1}d\gamma.$$

This implies the assertion by the uniqueness of the Laplace transform. Differentiating the formula $\nu_1([0,\frac{r^2}{2}]) = \nu_0([0,r])$, it follows easily that

$$\nu_0(d\rho) = \frac{2\pi^{\frac{d}{2}}}{\Gamma(\frac{d}{2})}\rho^{d-1}d\rho.$$

(2) Let $\mathbf{R}^d \setminus \{0\}$ be parametrized by $(u,\rho) \in S_{d-1} \times (0,+\infty)$, using the homeomorphism $x \mapsto (u,\rho) = (\frac{x}{\|x\|}, \|x\|)$. Since m is rotation invariant, it follows from Problem III-2 that $dm(u,\rho) = \sigma(du) \otimes \nu(d\rho)$, where σ is the uniform measure on the unit sphere S_{d-1} of $\mathbf{R}^d$. Setting $f_1(\rho) = \int f(u\rho)\sigma(du)$, it is clear that the image of $d\mu(u,\rho) = f(u,\rho)dm(u,\rho)$ under $(u,\rho) \mapsto \rho$ is $f_1(\rho)\nu_0(d\rho)$. Moreover, if μ is rotation invariant, Problem II-2 implies that $d\mu(u,\rho) = \sigma(du) \otimes f_1(\rho)\nu_0(d\rho)$. Since the measures $f_1(\|x\|)m(dx)$ and $f(x)m(dx)$ coincide, $f_1(\|x\|) = f(x)$ m-almost everywhere.

Problem III-4. *Euclidean space* $\mathbf{R}^d$ *is equipped with the scalar product* $\langle x,t\rangle$ *and the norm* $\|t\|$. Γ *is the usual Euler function.*

(1) Use Problem III-3 to evaluate $I = \int_{\mathbf{R}^d} \frac{dx}{(1+\|x\|^2)^{\frac{d+1}{2}}}$. *If* a *and* t *are in* $\mathbf{R}^d$ *and* $p > 0$, *use Problem II-18 to conclude that*

$$\frac{\Gamma(\frac{d+1}{2})}{\pi^{\frac{d+1}{2}}}\int_{\mathbf{R}^d} e^{i\langle x,t\rangle}\frac{p\,dx}{(p^2+\|x-a\|^2)^{\frac{d+1}{2}}} = e^{-p\|t\|+i\langle a,t\rangle}.$$

(2) Show that, if $x \in \mathbf{R}^d$ *and* $p > 0$,

$$2^d(\frac{d+1}{2})\pi^{\frac{d-1}{2}}\frac{p}{(p^2+\|x\|^2)^{\frac{d+1}{2}}} = \int_{\mathbf{R}^d} e^{-p\|t\|+i\langle x,t\rangle}dt.$$

SOLUTION. (1) By Problem III-3, the image ν_0 of Lebesgue measure dx on $\mathbf{R}^d$ under the mapping $x \mapsto \|x\|$ is $\nu_0(d\rho) = \frac{2\pi^{\frac{d}{2}}}{\Gamma(\frac{d}{2})}\rho^{d-1}d\rho$. Hence

$$I = \frac{2\pi^{\frac{d}{2}}}{\Gamma(\frac{d}{2})}\int_0^\infty \frac{\rho^{d-1}d\rho}{(1+\rho^2)^{\frac{d+1}{2}}} = \frac{\pi^{\frac{d}{2}}}{\Gamma(\frac{d}{2})}\int_0^\infty \frac{x^{\frac{d}{2}-1}dx}{(1+x)^{\frac{d+1}{2}}}.$$

This last integral is the usual Euler integral of the second kind $B(\frac{d}{2},\frac{1}{2})$ (see, for example, Problem IV-11), and hence

$$I = \frac{\pi^{\frac{d}{2}}}{\Gamma(\frac{d}{2})}\frac{\Gamma(\frac{d}{2})\Gamma(\frac{1}{2})}{\Gamma(\frac{d+1}{2})} = \frac{\pi^{\frac{d+1}{2}}}{\Gamma(\frac{d+1}{2})}.$$

The second integral of (1) now follows directly from Problem II-18.
(2) Set $a = 0$ in the preceding formula. Since $t \mapsto \exp(-p\|t\|)$ is an integrable function, the Fourier inversion theorem (III-2.4.6) gives the desired result.

Problem III-5. *Let k be a positive integer. In the Euclidean space $\mathbf{R}^{2k-1}$, the norm is written as $\|t\|$ and the scalar product as $\langle x, t\rangle$. Consider the map $\varphi : \mathbf{R}^{2k-1} \mapsto [0,1]$ defined by*

$$\varphi(t) = \left[(1 - \|t\|)^+\right]^k.$$

(1) Using Problem III-1, show that there exists a continuous function $f : [0, +\infty) \to \mathbf{R}$ such that

$$f(\|x\|) = \int_{\mathbf{R}^{2k-1}} \exp(i\langle x, t\rangle)\varphi(t)dt.$$

(2) Use Problems III-3 and III-4 to show that, for every $s > 0$,

$$I = \int_0^\infty e^{-su}u^{3k-1}f(u)du = C_k\left[\int_0^\infty e^{-su}(1 - \cos u)du\right]^k,$$

where C_k is a constant.
(3) Show that $f \geq 0$ and that $\int_{\mathbf{R}^{2k-1}} f(x)dx < \infty$ by using Problem II-19 and the sequence of functions $f_n : [0, +\infty) \to \mathbf{R}$ defined by $f_1(u) = 1 - \cos u$ and $f_{n+1}(u) = \int_0^u f_n(u - \rho)f_1(\rho)d\rho$.

Conclude that φ is the Fourier transform of a probability measure on $\mathbf{R}^{2k-1}$. Compute it for $k = 1$ and $k = 2$.
(4) Suppose that $g : [0, +\infty) \to \mathbf{R}$ is continuous and satisfies the following conditions: (i) $g(0) = 1$; (ii) $(-1)^{k-1}g^{(k-1)}(x)$ exists and is convex on $(0, +\infty)$; and (iii) $\lim_{x\to+\infty} g(x) = \lim_{x\to+\infty} g^{(k-1)}(x) = 0$. Use Problem II-6 to show that $g(\|t\|)$ is the Fourier transform of a probability measure on $\mathbf{R}^{2k-1}$.

SOLUTION. (1) $\varphi(t)dt$ defines a rotation-invariant measure on $\mathbf{R}^{2k-1}$. By Problem III-1, its Fourier transform (in x) depends only on $\|x\|$.
(2) We may consider $\varphi(t)dt$ as a measure on $\mathbf{R}^{2k-1} \setminus 0$ since it has no weight at the origin. Identifying $\mathbf{R}^{2k-1} \setminus 0$ with the product $S \times (0, +\infty)$ under the map $t \mapsto (\frac{t}{\|t\|}, \|t\|) = (v, \rho)$, the measure $\varphi(t)dt$ is the product measure $\sigma(dv) \otimes \nu(d\rho)$, where σ is the uniform probability measure on the unit sphere S and $\nu(d\rho) = K_1\rho^{2k-2}[(1 - \rho)^+]^k d\rho$, with $K_1 = \frac{2\pi^{k-\frac{1}{2}}}{\Gamma(k-\frac{1}{2})}$ by Problem III-3. Then

$$f(\|x\|) = K_1 \int_0^1 \rho^{2k-2}(1 - \rho)^k d\rho \int_S \exp(i\rho\langle x, v\rangle)\sigma(dv).$$

Next, set $u = \|x\| > 0$ and $r = \rho u$ in the preceding integral:

$$u^{3k-1} f(u) = K_1 \int_0^u r^{2k-2}(u-r)^k dr \int_S \exp(ir\langle \frac{x}{\|x\|}, v\rangle)\sigma(dv).$$

It should be noted that, since σ is rotation invariant, the last integral depends on r but not on x. From now on, let x_1 in S be fixed.

$$\begin{aligned} I &= \int_0^\infty e^{-su} du \int_0^u r^{2k-2}(u-r)^k dr \int_S \exp(ir\langle x_1, v\rangle)\sigma(dv) \\ &= K_1 \int_0^\infty e^{-sw} w^k dw \int_0^\infty r^{2k-2} e^{-rs} dr \int_S \exp(ir\langle x_1, v\rangle)\sigma(dv) \\ &= \int_0^\infty e^{-sw} w^k dw \int_{\mathbf{R}^{2k-1}} \exp(-s\|t\| + i\langle x_1, t\rangle) dt. \end{aligned}$$

The first inequality was obtained by the change of variable $w = u - r$, and the second by writing dt in $\mathbf{R}^{2k-1} \setminus \{0\}$ as the product measure $\sigma(dv) \otimes K_1 \rho^{2k-1} d\rho$ on $S \times (0, +\infty)$. The last integral was evaluated in Problem III-4 and equals $K_2 s(1+s^2)^{-k}$, since $\|x_1\| = 1$, with $K_2 = 2^{k-1}\Gamma(k)\pi^{k-1}$. Since $\int_0^\infty e^{-sw} w^k dw$ is equal to $s^{-k-1}\Gamma(k+1)$,

$$I = C_k \frac{1}{s^k}(1+s^2)^K \quad \text{with} \quad C_k = 2^{2k-1}\Gamma(k)\Gamma(k+1)\pi^{k-1}.$$

Then, since $\int_0^\infty e^{-su} \cos u du = \operatorname{Re} \int_0^\infty e^{-(s+i)u} du = \frac{s}{s^2+1}$, it follows that $\int_0^\infty e^{-su}(1 - \cos u) du = \frac{1}{s(1+s^2)}$, which completes the proof.
(3) The following two properties are easily seen by induction on n:

$$0 \leq f_n(u) \leq \frac{u^{n-1}}{(n-1)!} \quad \text{and} \quad \int_0^\infty e^{-su} f_n(u) du = \frac{1}{s^n(1+s^2)^n}.$$

The uniqueness of the Laplace transform implies that $u^{3k-1} f(u) = C_k f_k(u)$ for almost every $u \geq 0$. Since f and the f_k are continuous on $[0, +\infty)$, this equality holds for every $u > 0$ and $\lim_{u \to 0} u^{-3k+1} f_k(u) = \frac{f(0)}{C_k}$. The function f is thus nonnegative and $K_1 f(u) u^{2k-2} \leq \frac{K_1 C_k}{(k-1)!} u^{-2}$; hence $\int_{\mathbf{R}^{2k-1}} f(\|x\|) dx = \int_0^\infty K_1 f(u) u^{2k-2} du < \infty$. The Fourier inversion theorem is therefore applicable and

$$\varphi(t) = \frac{1}{(2\pi)^{2k-1}} \int_{\mathbf{R}^{2k-1}} \exp(i\langle x, t\rangle) f(\|x\|) dx,$$

which shows that $\varphi(t)$ is the Fourier transform of the positive measure $\frac{1}{(2\pi)^{2k-1}} f(\|x\|) dx$ on $\mathbf{R}^{2k-1}$. Since $\varphi(0) = 1$ it is a probability measure.

If $k = 1$, the measure is $4\pi \frac{1-\cos x}{x^2} dx$, with Fourier transform $(1 - \|t\|)^+$.

If $k = 2$, we first calculate $f_2(u) = \int_0^u (1 - \cos\rho)(1 - \cos(u-\rho)) d\rho = u - \frac{3}{2}\sin u + \frac{u}{2}\cos u$, and obtain $\frac{2}{\pi^2}(\|x\| - \frac{3}{2}\sin\|x\| + \frac{\|x\|}{2}\cos\|x\|) dx$ as a measure on $\mathbf{R}^3$ with Fourier transform $[(1 - \|t\|)^+]^2$.

(4) By Problem II-6, there exists a measure $\nu \geq 0$ on $(0, +\infty)$ such that $\nu([x, +\infty)) < +\infty$ for all $x > 0$ and

$$g(x) = \int_0^\infty \left[(1 - \frac{x}{u}^+\right]^k \nu(du) \quad \text{for all } x > 0.$$

Since g is continuous at 0 and $g(0) = 1$, it follows from monotone convergence that $\int_0^\infty \nu(du) = 1$. Hence $g(\|t\|) = \int_0^\infty \varphi(\frac{\|t\|}{u})\nu(du)$. Set $h(x) = \frac{C_k f(\|x\|)}{(2\pi)^{2k-1}}$. Thus, for $u > 0$,

$$\varphi\left(\frac{\|t\|}{u}\right) = \int_{\mathbf{R}^{2k-1}} \exp(i\langle t, \frac{x}{u}\rangle)h(x)dx = \int_{\mathbf{R}^{2k-1}} \exp(i\langle x, t\rangle)h(uy)u^{2k-1}dy.$$

Setting $h_1(y) = \int_0^\infty h(uy)u^{2k-1}\nu(du)$ gives

$$g(\|t\|) = \int_{\mathbf{R}^{2k-1}} \exp(i\langle y, t\rangle)h_1(y)dy,$$

which proves the desired result.

A more probabilistic method is to consider a random variable X on $\mathbf{R}^{2k-1}$ such that $\mathbf{E}[\exp i\langle X, t\rangle] = \varphi(t)$ and a random variable U on $(0, +\infty)$ with distribution ν independent of X. Then $\mathbf{E}[\exp i\langle \frac{X}{U}, t\rangle] = g(\|t\|)$.

REMARK. The result of (4) for $k = 1$ is due to G. Polya (1923), and the general case to R. Askey (1972).

Problem III-6.

Let $\mathbf{C}$ denote the complex numbers. A function $p : \mathbf{C} \to [0, +\infty)$ is called a seminorm *if*

(i) $$p(\lambda z) = |\lambda| p(z) \quad \textit{for all } \lambda \in \mathbf{R} \textit{ and } z \in \mathbf{C}$$

and

(ii) $$p(z_1 + z_2) \leq p(z_1) + p(z_2) \quad \textit{for all } z_1 \textit{ and } z_2 \textit{ in } \mathbf{C}.$$

(1) Let $p : \mathbf{C} \to [0, +\infty)$ satisfy (i). Prove the equivalence of the following properties.

a) p is a seminorm.

b) $\{z : p(z) \leq 1\}$ is a convex subset of $\mathbf{C} = \mathbf{R}^2$.

c) For all α_1, α_2, α_3 such that $\alpha_1 < \alpha_2 < \alpha_3$ and $\alpha_3 - \alpha_1 < \pi$,

(iii) $p(e^{i\alpha_3})\sin(\alpha_2 - \alpha_1) + p(e^{i\alpha_1})\sin(\alpha_3 - \alpha_2) - p(e^{i\alpha_2})\sin(\alpha_3 - \alpha_1) \geq 0.$

(2) Let μ be a bounded positive measure on $[0, \pi)$. Show that

(iv) $$p_\mu(x + iy) = \int_0^\pi |x \sin\alpha - y\cos\alpha| \mu(d\alpha)$$

defines a seminorm. Show that $p_\mu = p_{\mu_1}$ *implies* $\mu = \mu_1$.

METHOD. Observe that $p(e^{i\theta})$ is the convolution of μ and $|\sin\theta|$ in the group $\mathbf{R}/\pi\mathbf{Z}$. (See III-1.8.)

(3) Let $0 \le \alpha_1 < \alpha_2 < \dots \alpha_n < \pi$, *with the convention that* $\alpha_0 = \alpha_n - \pi$ *and* $\alpha_{n+1} = \alpha_1 + \pi$. *The matrices* $A = (a_{ij})_{i,j=1}^n$, $B = (b_{ij})_{i,j=1}^n$, *and* $D = (d_{ij})_{i,j=1}^n$ *are defined as follows:*

$a_{ij} = |\sin(\alpha_i - \alpha_j)|$ *for all* $i, j = 1, \dots, n$.

$b_{ii} = -\sin(\alpha_{i+1} - \alpha_{i-1})$, $b_{i,i+1} = \sin(\alpha_i - \alpha_{i-1})$ *(with the convention that* $b_{n,n+1} = b_{n,1}$*),* $b_{i,i-1} = \sin(\alpha_{i+1} - \alpha_i)$ *(with the convention that* $b_{1,0} = b_{1,n}$*) for* $i = 1, \dots, n$, *and* $b_{ij} = 0$ *otherwise.*

$d_{ii} = 2\sin(\alpha_{i+1} - \alpha_i)\sin(\alpha_i - \alpha_{i-1})$ *for* $i = 1, \dots, n$, *and* $d_{ij} = 0$ *otherwise.*

Verify that $AB = D$. *If* $\mu = \sum_{j=1}^n m_j \delta_{\alpha_j}$, *where* $m_j > 0$ *and* δ_{α_j} *is the Dirac measure at* α_j *for* $j = 1, 2, \dots, n$, *calculate* $p(e^{i\theta})$ *and verify that*

$$(v) \qquad [m_1, m_2, \dots, m_n]A = \left[p_\mu(e^{i\alpha_1}), p_\mu(e^{i\alpha_2}), \dots, p_\mu(e^{i\alpha_n})\right].$$

(4) If p is a seminorm, show that there exists a bounded positive measure μ *on* $[0, \pi)$ *such that* $p = p_\mu$.

METHOD. Let $T = \{\alpha_1, \dots, \alpha_n\}$ with $\alpha_0 = \alpha_n - \pi < 0 \le \alpha_1 < \dots < \alpha_n < \pi < \alpha_{n+1} = \alpha_1 + \pi$. Show that there exists a seminorm p_T such that, if $0 \le \lambda \le 1$ and $j = 1, \dots, n$,

$$(vi) \qquad p_T\left[\lambda e^{i\alpha_j} + (1-\lambda)e^{i\alpha_{j+1}}\right] = \lambda p_T\left[e^{i\alpha_j}\right] + (1-\lambda)p_T\left[e^{i\alpha_{j+1}}\right],$$

and show by using (3) that there exists a measure μ_T concentrated on T such that $p_T = p_{\mu_T}$.

Next, let $\alpha_j = \frac{(j-1)\pi}{n}$ and set $p_n = p_T$ and $\mu_n = \mu_T$. Show that $p = \lim_{n\to\infty} p_n$ and that there exists a bounded positive measure μ on $[0, \pi)$ such that μ_n converges vaguely to μ as $n \to \infty$.

SOLUTION. (1) The equivalence (a) $\Leftrightarrow$ (b) is classical. We prove that (a) $\Rightarrow$ (c). If p is a seminorm, we take $z_1 = \rho_1 e^{i\alpha_1}$ and $z_3 = \rho_3 e^{i\alpha_3}$, with ρ_1 and ρ_3 chosen so that $z_1 + z_3 = e^{i\alpha_2}$. It is clear geometrically that ρ_1 and ρ_3 must be positive. Moreover, solving the linear system which defines them gives

$$\rho_1 = \frac{\sin(\alpha_3 - \alpha_2)}{\sin(\alpha_3 - \alpha_1)} \quad \text{and} \quad \rho_3 = \frac{\sin(\alpha_2 - \alpha_1)}{\sin(\alpha_3 - \alpha_1)}.$$

Inequality (iii) follows from (i) and (ii). The proof that (c) $\Rightarrow$ (a) is similar.
(2) It is clear that (iv) defines a nonnegative function satisfying (i) and (ii). In order to prove the uniqueness of μ, we write

$$p_\mu(e^{i\theta}) = \int_0^\pi |\sin(\theta - \alpha)|\mu(d\alpha).$$

Hence, for every $n \in \mathbf{Z}$.

$$\int_0^\pi p_\mu(e^{i\theta})e^{2in\theta}d\theta = \int_0^\pi |\sin\theta|e^{2in\theta}d\theta \times \int_0^\pi e^{2in\theta}\mu(d\theta).$$

Since $\int_0^\pi |\sin\theta|e^{2i\theta}d\theta = \frac{2}{(1-4n^2)}$ is not zero for any $z \in \mathbf{Z}$, we conclude that $p = p_{\mu_1}$ implies that $\int_0^\pi e^{2in\theta}(\mu(d\theta) - \mu_1(d\theta)) = 0$ for every $n \in \mathbf{Z}$. By the uniqueness of the Fourier transform of a measure, $\mu = \mu_1$.
(3) Equality between matrices is verified in the standard way. Then $p_\mu(e^{i\theta}) = \sum_{j=1}^n |\sin(\theta - \alpha_j)|m_j$ and (v) follows.
(4) Equation (vi) implies that $\{z : p_T(s) \leq 1\}$ is a convex polygon with vertices $\pm\frac{1}{p(e^{i\alpha_j})}e^{i\alpha_j}$, $j = 1, 2, \ldots, n$. (The case of a nonzero z_0 such that $p(z_0) = 0$ is easy to treat directly, by finding $p = p_{m\delta_\alpha}$ with $e^{i\alpha} = \frac{z_0}{\|z_0\|}$.) By (1b), p_T is a seminorm. By its definition, if $\alpha_j \leq \theta \leq \alpha_{j+1}$, $j = 1, \ldots, n$, then

$$(vii) \qquad p_T(e^{i\theta}) = \frac{\sin(\alpha_{j+1} - \theta)p(e^{i\alpha_j}) + \sin(\theta - \alpha_j)p(e^{i\alpha_{j+1}})}{\sin(\alpha_{j+1} - \alpha_j)}.$$

Next, we show that there exists a positive measure μ_T concentrated on T such that $p_T = p_{\mu_T}$.

Let $m_1, \ldots, m_n$ be defined by

$$[m_1, \ldots, m_n]A = [p(e^{i\alpha_1}), \ldots, p(e^{i\alpha_n})].$$

This can be done since $AB = D$ by (3), and hence A is invertible. Thus

$$2\sin(\alpha_{j+1} - \alpha_j)\sin(\alpha_j - \alpha_{j-1})m_j = p(e^{i\alpha_{j+1}})\sin(\alpha_j - \alpha_{j-1}) + p(e^{i\alpha_{j-1}})\sin(\alpha_j - \alpha_{j-1}) - p(e^{i\alpha_j})\sin(\alpha_{j+1} - \alpha_{j-1}),$$

a quantity which is nonnegative by (iii). Hence $m_j \geq 0$. Let $\mu_T = \sum_{j=1}^n m_j\delta_{\alpha_j}$. We now show that $p_T(e^{i\theta}) = \sum_{i=1}^n m_i|\sin(\theta - \alpha_i)|$. If $\alpha_j \leq \theta\alpha_{j+1}$, it can be verified in the standard way that
(viii)

$$\sin(\theta - \alpha_i) = \frac{\sin(\alpha_{j+1} - \theta)|\sin(\alpha_j - \alpha_i)| + \sin(\theta - \alpha_j)|\sin(\alpha_{j+1} - \alpha_i)|}{\sin(\alpha_{j+1} - \alpha_j)}$$

for $j = 0, 1, \ldots, n$. Comparing (vii) and (viii) gives $p_T = p_{\mu_T}$.

Next, let p_n and μ_n be as defined above. Conditions (i) and (ii) imply that p is continous on $\mathbf{C}$, since

$$|p(z_1) - p(z_2)| \leq p(z_1 - z_2) \text{for all } z_1 \text{ and } z_2 \text{ in } \mathbf{C}.$$

Hence $p(e^{i\theta})$ is uniformly continuous on $[0, \pi)$. If $j_n^{(\theta)}$ is the integer defined by $\alpha \leq \theta < \alpha_{j_n+1}$, it follows that for every $\epsilon > 0$ there exists $N(\epsilon)$ such that $n \geq N(\epsilon)$ implies $|p(e^{i\alpha_j}) - p(e^{i\theta})| \leq \epsilon$ for $j = j_n$ and $j = j_{n+1}$. We evaluate the difference $p_n(e^{i\theta}) - p(e^{i\theta})$, using (vii) and taking $j = j_n$; for simplicity, we set $a = \alpha_{j_n+1} - \theta$ and $b = \theta - \alpha_{j_n}$. Then

$$\begin{aligned} p_n(e^{i\theta}) - p(e^{i\theta}) &= \frac{\sin a + \sin b - \sin(a+b)}{\sin(a+b)} p(e^{i\theta}) \\ &\quad + \frac{\sin a}{\sin(a+b)} \left[p(e^{i\alpha_{j_n}}) - p(e^{i\theta})\right] \\ &\quad + \frac{\sin b}{\sin(a+b)} \left[p(e^{i\alpha_{j_n+1}}) - p(e^{i\theta})\right] \end{aligned}$$

Since it is clear that $\frac{(\sin a + \sin b) - \sin(a+b)}{\sin(a+b)}$ tends (uniformly in θ) to 0 as $n \to +\infty$, we conclude that $p_n(e^{i\theta}) \to p(e^{i\theta})$ uniformly.

Hence $\lim_{n\to\infty} \int_0^{\pi} e^{2ik\theta} p_n(e^{i\theta}) d\theta = \int_0^{\pi} e^{2ik\theta} p(e^{i\theta}) d\theta$ for every k in $\mathbf{Z}$. This implies that $\lim_{n\to\infty} \int_0^{2\pi} e^{2ik\theta} \mu_n(d\theta)$ exists and, by the theorem of Paul Lévy (III-3.1.2), μ_n converges vaguely to a bounded positive measure on $[0, \pi)$ such that

$$\lim_{n\to\infty} \int_0^{2\pi} e^{2ik\theta} \mu_n(d\theta) = \int_0^{2\pi} e^{2ik\theta} \mu(d\theta).$$

Hence

$$\int_0^{2\pi} e^{2ik\theta} p(e^{i\theta}) d\theta = \frac{2}{1-4n^2} \int_0^{2\pi} e^{2ik\theta} \mu(d\theta),$$

which is equivalent to $p = p_\mu$.

REMARKS. 1. A consequence of (4) is that every seminorm on $\mathbf{R}^2$ can be approximated by finite sums of the type $\sum_j |a_j x + b_j y|$, and not only by $\sup_j |a_j x + b_j y|$. For $\mathbf{R}^n$ with $n > 2$ this is false; in general, a seminorm can be approximated only by suprema of absolute values of linear functionals.
2. The relation described in the problem,

compact symmetric convex sets $\leftrightarrow$ positive measures on $\mathbf{R}/\pi\mathbf{Z}$,

can in fact be extended to the nonsymmetric case in the following way. Let $\mathcal{C}$ be the set of nonempty compact convex subsets of $\mathbf{R}^2$, let $\sim$ be the equivalence relation defined on $\mathcal{C}$ by "$C_1 \sim C_2$ if there exists (a, b) in $\mathbf{R}^2$ such that $C_1 + (a, b) = C_2$", and let $\mathcal{M}_0$ be the set of positive measures s on the group $\mathbf{R}/2\pi\mathbf{Z}$ such that $\int_0^{2\pi} e^{i\theta} s(d\theta) = 0$. Then there is a bijection between $\mathcal{C}/\sim$ and $\mathcal{M}_0$. To show this, set $h(\theta) = \max\{x\cos\theta + y\sin\theta : (x, y) \in C\}$ for $C \in \mathcal{C}$ and check that

a) the left derivative h'_- exists everywhere;

b) $s_g(\theta) = \int_0^{\theta} h(\alpha) d\alpha + h'_g(\theta)$ is nondecreasing;

c) the measure s_C on $\mathbf{R}/2\pi\mathbf{Z}$ defined by $s_C([\alpha, \beta]) = s_g(\beta) - s_g(\alpha)$, when α and β are points of continuity of s_g, is in $\mathcal{M}_0$ and satisfies $s_{C_1} = s_{C_2} \Leftrightarrow C_1 \sim C_2$; and

d) the map $\mathcal{C} \to \mathcal{M}_0$, $C \mapsto s_C$ is surjective.

If $\tilde{C}$ is the equivalence class of C, then $\tilde{C} \mapsto s_C$ is the desired bijection. This correspondence allows many properties of $\mathcal{C}$ to be expressed in terms of positive measures.

Problem III-7. *Let* $\mathbf{C}$ *be the set of complex numbers, identified with* $\mathbf{R}^2$*, and let* p *be a seminorm on* $\mathbf{C}$*. Show that* $\exp(-p(t))$ *is the Fourier transform of a probability measure on* $\mathbf{R}^2$*.*

METHOD. Use the fact, proved in Problem III-6, that there exists a sequence of measures $\mu_n \geq 0$ on $[0, \pi)$, concentrated at a finite number of points, such that

$$p(x+iy) = \lim_{n\to\infty} \int_0^{\pi} |x \sin\alpha - y\cos\alpha| \mu_n(d\alpha).$$

Also use the formula $\mathrm{e}^{-|t|} = \int_{-\infty}^{+\infty} \mathrm{e}^{itx} \frac{dx}{\pi(1+x^2)}$, which appeared in Problem III-4.

SOLUTION. Set $\mu_n = \sum_{j=1}^{N_n} m_j \mathrm{e}^{i\alpha_j}$ with $0 \leq \alpha_1 < \ldots < \alpha_{N_n} < \pi$ and positive m_j. Then, if $p_n(x+iy) = \int_0^{\pi} |x\sin\alpha - y\cos\alpha|\mu_n(d\alpha)$ and $t = (t_1, t_2)$, we can write

$$\exp(-p_n(t)) = \prod_{j=1}^{N_n} \exp(-m_j|t_1 \sin\alpha_j - t_2\cos\alpha_j|).$$

Next, consider a finite sequence of independent real-valued random variables $\{X_j\}_{j=1}^{N_n}$ such that X_j has density $\frac{m_j}{\pi(m_j^2+x^2)}$. Let Z_j be the random variable of $\mathbf{C} = \mathbf{R}^2$ defined by $Z_j = -iX_j\mathrm{e}^{i\alpha_j}$ and let $t = (t_1, t_2)$. Then $\mathbf{E}[\exp(i\langle Z_j, t\rangle)] = \exp(-m_j|t_1 \sin\alpha_j - t_2 \cos\alpha_j)$ and thus $\mathbf{E}[\exp(i\langle \sum Z_j, t\rangle)] = \exp(-p_n(t))$. This shows that $\exp(-p_n(t))$ is the Fourier transform of a probability measure on $\mathbf{R}^2$. Since $\exp(-p_n(t)) \to \exp(-p(t))$ as $n \to \infty$ and $\exp(-p(t)$ is continuous, Paul Lévy's theorem gives the desired result.

REMARKS. This result is due to T. Ferguson (1962). It is false in higher dimensions: only for certain norms (like the Euclidean norm) is $\exp(-p(t))$ the Fourier transform of a probability measure. See Problem III-8 for a counterexample.

Problem III-8. *(1) What is the image* ν *in* $\mathbf{R}$*, under the projection* $(x_0, \ldots, x_n) \mapsto x_0$*, of the measure* $\exp(-\max_{j=0,\ldots,n} |x_j|)dx_0 dx_1 \ldots dx_n$ *in* $\mathbf{R}^n$*? (See Problem I-14.)*
(2) Compute the Fourier transform of ν*.*
METHOD. Show that $k!(1-it)^{-(k+1)} = \int_0^{\infty} x^k \exp(-x+itx)dx$ for t real and k a nonnegative integer.

(3) Conclude that $\varphi_{n+1}(t) = \exp(-\max_{j=0,\dots,n}|t_j|)$ *is not the Fourier transform of a probability measure on* $\mathbf{R}^{n+1}$ *when* $n \geq 2$.

SOLUTION. (1) The density of the measure ν is

$$f(x_0) = \int_{\mathbf{R}^n} \exp(-\max_{j=0,\dots,n}|x_j|)dx_1\dots dx_n.$$

Considering separately the cases $\max_{j=1,\dots,m}|x_j| \leq |x_0|$ and $\max_{j=1,\dots,m}$ $|x_j| > |x_0|$, we find that

$$f(x_0) = 2^n|x_0|^n \mathrm{e}^{-|x_0|} + n2^n \int_0^\infty \mathrm{e}^{-x_1}x_1^{n-1}dx_1.$$

Setting $x_1 = x_0| + u$ in the last integral gives

$$f(x_0) = 2^n \mathrm{e}^{-|x_0|}\left(|x_0|^n + n\int_0^\infty \mathrm{e}^{-u}(u+|x_0|)^{n-1}du\right)$$

and, by a standard calculation,

$$f(x_0) = 2^n n! \mathrm{e}^{-|x_0|}\sum_{k=0}^n \frac{|x_0|^k}{k!}.$$

(2) Since $\int_0^\infty \exp(-x+itx)dx = (1-it)^{-1}$, the indicated equality is an immediate result of differentiating under the integral sign, which is legitimate because $x^k\mathrm{e}^{-x}$ is integrable on $[0,+\infty)$. Since f is an even function, we may write

$$\widehat{\nu}(t) = \int_{-\infty}^{+\infty} \mathrm{e}^{itx}f(x)dx = 2\mathrm{Re}\int_0^\infty \mathrm{e}^{itx}f(x)dx = 2^{n+1}n!\mathrm{Re}\sum_{k=0}^n (1-it)^{-k-1}.$$

Hence

$$\begin{aligned}\widehat{\nu}(t) &= \tfrac{2^{n+1}n!}{t}\mathrm{Im}(1-it)^{-n-1} = \tfrac{2^{n+1}n!}{(1+t^2)^{n+1}}\tfrac{1}{t}\mathrm{Im}(1+it)^{n+1} \\ &= \tfrac{2^{n+1}n!}{(1+t^2)^{n+1}}\textstyle\sum_{0\leq k\leq \frac{n}{2}} C_{n+1}^{2k+1}(-t^2)^k.\end{aligned}$$

In particular,

$$\begin{aligned}&\text{if } n=1 \qquad & \widehat{\nu}(t) &= \tfrac{4}{(1+t^2)^2}\\ &\text{if } n=2 & \widehat{\nu}(t) &= \tfrac{16}{(1+t^2)^3}(3-t^2)\\ &\text{if } n=3 & \widehat{\nu}(t) &= \tfrac{2^6 3}{(1+t^2)^4}(1-t^2).\end{aligned}$$

(3) We show that $\varphi_3(t)$ is not the Fourier transform of a probability measure on $\mathbf{R}^3$. If it were, since φ_3 is integrable the density of the probability measure would be

$$f_3(x_0,x_1,x_2) = \frac{1}{(2\pi)^3}\int_{\mathbf{R}^3} \varphi_3(t_0,t_1,t_2)\exp[i(t_0x_0+t_1x_1+t_2x_2)]dt_0dt_1dt_2$$

by the Fourier inversion theorem, and f_3 would be *continuous* and positive. But, by the calculation above,

$$f_3(x_0,0,0) = \frac{2(3-x_0^2)}{\pi^3(1+x_0^2)^3} < 0 \quad \text{if } |x_0| > \sqrt{3}.$$

Since f_3 is continuous, $f_3 dx_0 dx_1 dx_2$ cannot be a measure ≥ 0.

If $n > 3$, note that

$$\varphi_3(t_0,t_1,t_2) = \varphi_{n+1}(t_0,t_1,t_2,0,0,\dots,0).$$

If φ_{n+1} is the Fourier transform of the density f_{n+1}, then

$$\int_{\mathbf{R}^{n+1-3}} f_{n+1}(x_0,x_1,x_2,\dots x_n)dx_3\dots dx_n = f_3(x_0,x_1,x_2),$$

and f_{n+1} cannot be positive.

REMARK. (3) is due to C. Herz (1963).

Problem III-9. *Let E be n-dimensional Euclidean space.*
(1) If $\alpha > 0$, $\beta > 0$, and $\alpha+\beta < n$, show that there exists a constant $K(\alpha,\beta)$ such that $I(y) = \int_E \|x\|^{\alpha-n}\|y-x\|^{\beta-n}dx = K(\alpha,\beta)\|y\|^{\alpha+\beta-n}$.
METHOD. Use Problem III-3.

(2) Let $0 < \gamma < n$ and let M_γ be the set of positive measures μ, not necessarily bounded, such that $f(\mu) = \int_{E\times E}\|x-y\|^{\gamma-n}\mu(dx)\mu(dy) < \infty$. Show that, if μ and ν are in M_γ,

$$\left|\int_{E\times E}\|x-y\|^{\gamma-n}\mu(dx)\nu(dy)\right| \leq \sqrt{f(\mu)f(\nu)}.$$

SOLUTION. (1) The function $g : x \to \|x\|^{\alpha-n}\|y-x\|^{\beta-n}$ is integrable. For there exist constants A_y and B_y such that

$$0 \leq g(x) \leq A_y\|x\|^{\alpha-n} \quad \text{if } \|x\| \leq 1$$

and

$$0 \leq g(x) \leq B_y\|x\|^{\alpha+\beta-2n} \quad \text{if } \|x\| \geq 1.$$

But, by Problem III-3, if $k_n = \frac{2\pi^{\frac{n}{2}}}{\Gamma(\frac{n}{2})}$,

$$\int_{\{x:\|x\|\leq 1\}} \|x\|^{\alpha-n}dx = k_n\int_0^1 \rho^{\alpha-1}d\rho = \frac{k_n}{\alpha} < \infty$$

and

$$\int_{\{x:\|x\|\geq 1\}} \|x\|^{\alpha+\beta-2n} dx = k_n \int_1^\infty \rho^{\alpha+\beta-n-1} d\rho = \frac{k_n}{n-\alpha-\beta} < \infty.$$

Next, making the change of variable $x = \|y\|u$ in the integral I, we obtain $I(y) = I(\frac{y}{\|y\|})\|y\|^{\alpha+\beta-n}$. The next step is to show that $I(s)$ is a constant when s ranges over the unit sphere of E. If $\|s_1\| = \|s\| = 1$ there exists a rotation g of E such that $g(s) = s_1$. Making the change of variable $u = g(v)$ in the integral $I(s_1) = \int_E \|u\|^{\alpha-n}\|s_1-u\|^{\beta-n} du$ gives $\|g(s)-g(v)\| = \|s-v\|$ and $du = (\det g)dv = dv$. Hence $I(s_1) = I(s) = K(\alpha, \beta)$.
(2) Fubini's theorem implies that if μ is in M_α, then $H_\mu^\alpha(z) = \int_E \|x - z\|^{\alpha-n}\mu(dx)$ is measurable and μ-integrable. Since H_μ^α is also positive we may write, for $\alpha > 0$, $\beta > 0$, and $\alpha + \beta < n$,

$$\begin{aligned}\int_E H_\mu^\alpha(z) H_\nu^\beta(z) dz &= \int_E \int_E \mu(dx)\nu(dy) \int_E \|x-z\|^{\alpha-n}\|y-z\|^{\beta-n} dz \\ &= K(\alpha,\beta) \int_{E\times E} \|x-y\|^{\alpha+\beta-n} \mu(dx)\nu(dy).\end{aligned}$$

This follows from (1), and is valid whether or not the integrals converge.
Next, setting $\alpha = \beta = \frac{\gamma}{2}$ and using Schwarz's inequality,

$$\left| \int_E H_\mu^{\frac{\gamma}{2}}(z) H_\nu^{\frac{\gamma}{2}}(z) dz \right|^2 \leq \left[\int_E (H_\mu^{\frac{\gamma}{2}}(z))^2 dz \right] \left[\int_E (H_\nu^{\frac{\gamma}{2}}(z))^2 dz \right].$$

Hence, applying the preceding inequality to the three pairs (μ, ν), (μ, μ), and (ν, ν),

$$\left| \frac{1}{K(\alpha,\beta)} \int_{E\times E} \|x-y\|^{\alpha+\beta-n} \mu(dx)\nu(dy) \right|^2 \leq \frac{1}{K(\alpha,\beta)} f(\mu) \times \frac{1}{K(\alpha,\beta)} f(\nu),$$

which is the desired inequality.

Problem III-10. *Let M be the space of* real *Radon measures on $\mathbf{U} = \{z : z \in \mathbf{C}$ and $|z| = 1\}$ and let F^+ (resp. F^-) be the vector space over $\mathbf{R}$ of complex functions defined in $\{z : |z| > 1\} = D^+$ (resp. in $\{z : |z| < 1\} = D^-$). For $\mu \in M$, we define*

$$f_\mu^+(z) = \int_{\mathbf{U}} (e^{i\theta} - z)^{-1} d\mu(e^{i\theta}) \quad \textit{for } z \in D^+$$

$$f_\mu^-(z) = \int_{\mathbf{U}} (e^{i\theta} - z)^{-1} d\mu(e^{i\theta}) \quad \textit{for } z \in D^-$$

(1) Show that the linear mapping $\mu \mapsto f_\mu^+$ from M to F^+ is injective.
METHOD. Expand f^+ in a power series in $\frac{1}{z}$.

(2) Find the kernel of the linear mapping $\mu \mapsto f_\mu^-$ from M to F^-.

SOLUTION. (1) If $z \in D^+$, then

$$f_\mu^+ = -\frac{1}{z}\int_{\mathbf{U}}(1-\frac{e^{i\theta}}{z})^{-1}d\mu(e^{i\theta}) = -\sum_{n=0}^{\infty}\frac{1}{z^{n+1}}\widehat{\mu}(n),$$

with $\widehat{\mu}(n) = \int_{\mathbf{U}} e^{in\theta}d\mu(e^{i\theta})$. Reversing the orders of the integral $\int_{\mathbf{U}}$ and the sum $\sum_{n=0}^{\infty}$ is easily justified here, since $\sum_{n=0}^{\infty}\frac{1}{|z|^{n+1}} < \infty$ and $|\widehat{\mu}(n)|$ is bounded by $\|\mu\|$. Now, if $f_\mu^+(z) = 0$ for all $z \in D^+$, then $\widehat{\mu}(n) = 0$ for all $n \in \mathbf{Z}$. By the uniqueness of the Fourier transform, $\mu = 0$.
(2) If $z \in D^-$, then

$$f_\mu^-(z) = \int_{\mathbf{U}}(1-e^{-i\theta}z)^{-1}e^{-i\theta}d\mu(e^{i\theta}) = \sum_{n=0}^{\infty} z^n\widehat{\mu}(-n-1),$$

with justifications analogous to those above. Then $\widehat{\mu}(n) = 0$ for all $n \neq 0$ when $f^-(z) = 0$ for all $z \in D^-$, and measures of this type have the form

$$d\mu_\lambda(e^{i\theta}) = \lambda d\theta,$$

where λ is an arbitrary real constant. Conversely, it is clear that $f_{\mu_\lambda} = 0$ and that $\{\mu_\lambda : \lambda \in \mathbf{R}\}$ is the kernel of $\mu \mapsto f_\mu^-$.

REMARKS. 1. Although f_μ^+ determines μ, f_μ^- does not.
2. The situation is completely different if μ is complex, since there exist complex measures, like $d\mu(e^{i\theta}) = e^{-i\theta}d\theta$, for which $\widehat{\mu}(n) = 0$ for all $n \geq 0$.

Problem III-11. *Let $P(x_1, \ldots, x_n) = P(x)$ be a homogeneous polynomial of degree m in n variables which is harmonic; that is, $\sum_{k=1}^{n}\frac{\partial^2 P}{\partial x_k^2}(x) = 0$ for all x in $\mathbf{R}^n$. For a fixed $\sigma < 0$, let*

$$f(x) = (\sigma\sqrt{2\pi})^{-n}\exp\left(\frac{-\|x\|^2}{2\sigma^2}\right)P(x), \quad \textit{with} \quad \|x\|^2 = \textstyle\sum_{k=1}^{n} x_k^2.$$

Show by induction on m that there exists a number $K_m(\sigma)$ such that

$$\widehat{f}(t) = K_m(\sigma)P(t)\exp\left(\frac{-\sigma^2\|t\|^2}{2}\right).$$

METHOD. $mP = \sum_k x_k \frac{\partial P}{\partial x_k}$.

SOLUTION. Let $f_k(x) = (\sigma\sqrt{2\pi})^{-n}\exp(\frac{-\|x\|^2}{2\sigma^2})\frac{\partial P}{\partial x_k}(x)$ and let $g_k(x) = x_k f_k(x)$. The result is true for $m = 0$. Assume that it holds for $m-1$ and observe that $\frac{\partial P}{\partial x_k}$ is homogeneous of degree $m-1$ and harmonic. Then

$$\begin{aligned}\widehat{f}(t) &= \tfrac{1}{m}\textstyle\sum_{k=1}^n \widehat{g}_k(t) = \tfrac{1}{im}\sum_{k=1}^n \frac{\partial}{\partial t_k}\widehat{f}_k(t)\\ &= \tfrac{K_{m-1}(\sigma)}{im}\textstyle\sum_{k=1}^n \frac{\partial}{\partial t_k}[\frac{\partial P}{\partial t_k}(t)\exp(\frac{-\sigma^2\|t\|^2}{2})]\\ &= -\sigma^2\tfrac{K_{m-1}(\sigma)}{im}\exp(\frac{-\sigma^2\|t\|^2}{2})\textstyle\sum_{k=1}^n \frac{\partial P}{\partial t_k}(t)\\ &= -\sigma^2\tfrac{K_{m-1}(\sigma)}{i}\exp(\frac{-\sigma^2\|t\|^2}{2})P(t).\end{aligned}$$

The first and the last equality use homogeneity, the second a legitimate differentiation under the summation sign, and the fourth the harmonicity of P. Thus $K_m(\sigma) = K_{m-1}(\sigma)(i\sigma^2) = (i\sigma^2)^m K_0(\sigma) = (i\sigma^2)^m$.

Problem III-12. *The goal of this problem is to prove the following inequality of S. Bernstein: If μ is a complex measure on $[-a,+a]$, then $|\widehat{\mu}'(t)| \le a\sup_{s\in\mathbf{R}}|\widehat{\mu}(s)|$.*
(1) Consider the odd *function $h(\theta)$ of period 2π defined by $h(\theta) = \theta$ if $0 \le \theta \le \frac{\pi}{2}$ and $h(\theta) = \pi - \theta$ if $\frac{\pi}{2} \le \theta \le \pi$.*
(a) Compute $\nu_n = (2i\pi)^{-1}\int_{-\pi}^{+\pi} h(\theta)\exp(-in\theta)d\theta$ for n in $\mathbf{Z}$.
(b) If ν is the measure defined on $\mathbf{R}$ by $\nu = \sum_{n=-\infty}^{\infty}\nu_n\delta_n$, where δ_n is the Dirac measure at n, show that ν is bounded and that $h(\theta) = i\int_{-\infty}^{+\infty}\exp(ix\theta)\,\nu(dx)$.
(2) If μ is a complex measure on $[-\frac{\pi}{2},\frac{\pi}{2}]$, let

$$f(t) = \int_{-\frac{\pi}{2}}^{\frac{\pi}{2}} \exp(it\theta)\mu(d\theta).$$

(a) Show that $f(t) = (f * \nu)(t)$ for all real t.
(b) If $\mu = \mu_0 = (2i)^{-1}(\delta_{\frac{\pi}{2}} - \delta_{-\frac{\pi}{2}})$, deduce from (a) that $\sum_{k=-\infty}^{+\infty}(2k-1)^{-2} = \frac{\pi^2}{4}$.
(c) Returning to the general case, deduce from (a) and (b) that

$$|f'(t)| \le \frac{\pi}{2}\sup_{s\in\mathbf{R}}|f(s)| \quad \textit{for all } t \textit{ in } \mathbf{R}.$$

Show that equality holds if and only if μ is concentrated at the points $\pm\frac{\pi}{2}$.
(3) Prove Bernstein's inequality and discuss in detail the case of equality.

SOLUTION. (1) The computation of ν_n is standard. For k an integer,

$$\nu_{2k} = 0 \quad\text{and}\quad \nu_{2k-1} = \frac{2}{\pi}\cdot\frac{(-1)^k}{(2k-1)^2}.$$

That $\sum_n |\nu_n| < \infty$ is clear. It follows that

$$\theta \mapsto i\sum_n \nu_n e^{in\theta} = i\int_{-\infty}^{+\infty} e^{ix\theta}\nu(dx)$$

is continuous, has period 2π, and, by the uniqueness of the Fourier series expansion, coincides with $h(\theta)$.

(2a) Since differentiation under the integral sign is clearly permissible,

$$\begin{aligned} f'(t) &= \int_{-\frac{\pi}{2}}^{+\frac{\pi}{2}} e^{i\theta t}(i\theta)\mu(d\theta) = \int_{-\frac{\pi}{2}}^{+\frac{\pi}{2}} e^{i\theta t}(ih(\theta))\mu(d\theta) \\ &= \int_{-\frac{\pi}{2}}^{+\frac{\pi}{2}} e^{i\theta t}\left(\int_{-\infty}^{+\infty} e^{-ix\theta}\nu(dx)\mu(d\theta)\right) = \int_{-\infty}^{+\infty} f(t-z)\nu(dx) \\ &= \tfrac{2}{\pi}\textstyle\sum_k \frac{(-1)^k}{(2k-1)^2} f(t-(2k-1)), \end{aligned}$$

where the next-to-last equality follows from Fubini.

b) If $\mu = \mu_0$, then $f(t) = \sin\frac{\pi}{2}t$, $f(t-(2k-1)) = (-1)^k\cos\frac{\pi}{2}t$, and $f'(t) = \frac{\pi}{2}\cos\frac{\pi}{2}t$, which gives the result.

c) By (a) and (b),

$$|f'(t)| \le \frac{2}{\pi}\sum_k |f(t-2k+2)|(2k-1)^{-2} \le \|f\|_\infty \frac{2}{\pi}\sum_k (2k-1)^{-2} = \frac{\pi}{2}\|f\|_\infty.$$

The most delicate point is the case of equality. Suppose that there exists $t_0 \in \mathbf{R}$ such that $|f'(t_0)| = \frac{\pi}{2}\|f\|_\infty$. Then

$$0 = \frac{\pi}{2}\|f\|_\infty - |f'(t_0)| \ge \frac{2}{\pi}\sum_k (\|f\|_\infty - |f(t_0-2k+1)|)(2k-1)^{-2} \ge 0,$$

and hence $|f(t_0-2k+1)| = \|f\|_\infty$ for every integer k. Introducing real numbers α_k and β such that

$$f(t_0-2k+1) = \|f\|_\infty \exp(i\alpha_k) \quad \text{and} \quad f'(t_0) = \frac{\pi}{2}\|f\|_\infty \exp(i\beta),$$

we obtain

$$\begin{aligned} \tfrac{\pi}{2}\|f\|_\infty &= f'(t_0)\exp(-i\beta) = \frac{2}{\pi}\sum_k \|f\|_\infty \exp[i(\alpha_k-\beta)](2k-1)^{-2}(-1)^k \\ 1 &= \frac{4}{\pi^2}\sum_k \exp[i(\alpha_k-\beta)](2k-1)^{-2}(-1)^k \\ 0 &= \frac{4}{\pi^2}\sum_k (2k-1)^{-2}(1-\exp[i(\alpha_k-\beta+k\pi)]) \\ 0 &= \sum_k (2k-1)^{-2}(1-\cos(\alpha_k-\beta+k\pi)). \end{aligned}$$

Hence, for every integer k,

$$\cos(\alpha_k-\beta+k\pi) = 1, \quad \exp(i\alpha_k) = (-1)^k\exp(i\beta), \quad \text{and}$$

$$\int_{-\frac{\pi}{2}}^{+\frac{\pi}{2}} \exp[i(t_0+1)\theta - 2ik\theta]\mu(d\theta) = (-1)^k\|f\|_\infty \exp(i\beta).$$

Now let μ_1 be the measure on $\mathbf{R}$ defined by $\mu_1(d\theta) = \exp\{i[(t_0+1)\theta - \beta]\}\|f\|_\infty^{-1}\mu(d\theta)$. It satisfies

$$\int_{-\frac{\pi}{2}}^{+\frac{\pi}{2}} \exp(-2ik\theta)\mu_1(d\theta) = (-1)^k \quad \text{for every integer } k.$$

Next, consider the image $\tilde{\mu}_1$ of μ_1 under the canonical mapping of $\mathbf{R}$ into $\mathbf{R}/\pi\mathbf{Z}$. Then, since $\int_{\mathbf{R}/\pi\mathbf{Z}} \exp(-2ik\theta)\tilde{\mu}_1(d\theta) = (-1)^k$, it follows from the uniqueness of the Fourier transform in the group $\mathbf{R}/\pi\mathbf{Z}$ that $\tilde{\mu}_1$ must be the Dirac measure at $\frac{\pi}{2}$ (which is identified here with $-\frac{\pi}{2}$). Hence μ_1, *as a measure on* $\mathbf{R}$, is concentrated at the two points $-\frac{\pi}{2}$ and $\frac{\pi}{2}$, and so is μ. Conversely, it remains to verify that if $\mu = a\delta_{-\frac{\pi}{2}} + b\delta_{\frac{\pi}{2}}$ with a and b complex, then $\|f\|_\infty = |a| + |b|$. Since f' is of the same form as f, this will imply that $\|f'\|_\infty = \frac{\pi}{2}(|a|+|b|) = \frac{\pi}{2}\|f\|_\infty$. It suffices to write $a = |a|\exp(i(m+c))$ and $b = |b|\exp(i(m-c))$. Then $f(-\frac{2c}{\pi}) = (|a|+|b|)\exp(im)$, and hence $\|f\|_\infty \geq |a| + |b|$. Since $\|f\|_\infty \leq |a| + |b|$ trivially, equality will hold precisely for measures concentrated on $\{-\frac{\pi}{2}, +\frac{\pi}{2}\}$.
(4) The general case follows immediately from the case $a = \frac{\pi}{2}$ just considered. Equality corresponds to measures concentrated on $\{-a, +a\}$.

Problem III-13. *Let* $f : (0, +\infty) \to \mathbf{R}$ *be measurable and satisfy*

$$f(x+y) = f(x) + f(y) \quad \textit{for all } x \textit{ and } y > 0.$$

(1) If $\varphi(t) = \int_0^1 \exp[itf(x)]dx$ *for* $t \in \mathbf{R}$, *show that* $y \mapsto \varphi(t)\exp[itf(y)]$ *is continuous on* $(0, +\infty)$ *and conclude that* f *is continuous.*
(2) Show that $f(x) = xf(1)$ *for* $x > 0$.

SOLUTION. (1) Since $|\exp[itf(x)]| = 1$ is measurable, $\varphi(t)$ exists. Hence

$$\varphi(t)\exp[itf(y)] = \int_0^1 \exp[itf(x+y)]dx = \int_y^{y+1} \exp[itf(u)]du.$$

Since $y \mapsto \int_0^y \exp[itf(u)]du$ is continuous, $y \mapsto \exp(itf(y))$ is continuous for all real t. Since $\exp(ita) = 1$ for all $t \in \mathbf{R}$ implies $a = 0$, we conclude that f is continuous. (2) If p and q are positive integers, it is easy to see by induction on p that $f(px) = p(fx)$; hence $f(\frac{1}{q}) = \frac{1}{q}f(1)$ and $f(\frac{p}{q}) = \frac{p}{q}f(1)$. Since the rationalsare dense in $(0, +\infty)$, it follows that $f(x) = xf(1)$ for all x.

Problem III-14. *Let* E *be a real vector space of finite dimension* n *and let* $\widehat{E}$ *be its dual. Let* $e_1, \ldots, e_n$ *be a basis of* E. *The dual basis* $e_1^*, \ldots, e_n^*$ *of* $\widehat{E}$ *is defined by* $\langle e_j, e_i^* \rangle = 0$ *if* $j \neq i$ *and* 1 *if* $j = i$, *where* $\langle\ ,\ \rangle$ *is the canonical bilinear form on* $E \times \widehat{E}$. E *and* $\widehat{E}$ *are equipped with Lebesgue measures* dx

and dt, respectively, such that, if $f \in L^1(E, dx)$ implies $\widehat{f} \in L^1(\widehat{E})$, where $\widehat{f}(t) = \int_E \exp(i\langle x,t\rangle) f(x)dx$, then $f(x) = (2\pi)^{-n} \int_{\widehat{E}} \exp(-i\langle x,t\rangle)\widehat{f}(t)dt$. Let Z denote the set of points $z = \sum_{i=1}^n z_i e_i$ of E such that the z_i are integers and let Z^* denote the set of points $\zeta = \sum_{i=1}^n \zeta_i e_i^*$ of $\widehat{E}$ such that the ζ_i are integers.
Prove Poisson's formula:

> *If f is in the space $\mathcal{S}$ of infinitely differentiable functions of rapid decrease, then for every t in $\widehat{E}$*
>
> $$\sum_{\zeta \in Z^*} \widehat{f}(2\pi\zeta + t) = [vol(e_1^*, \dots, e_n^*)]^{-1} \sum_{z \in Z} f(z) e^{i\langle z,t\rangle}.$$

METHOD. Show that $\sum_{z\in Z} |f(z)| < \infty$ and use Theorem III-4.2 to see that the left-hand side $\psi(t)$ of the equation exists. Observing that the set of periods of ψ contains $2\pi Z^*$, compute the Fourier coefficients of ψ.

SOLUTION. If $z = \sum_{i=1}^n z_i e_i$ and $|z| = (\sum_m |z_i|^2)^{\frac{1}{2}}$, we know that $\lim_{|z|\to\infty} (1 + |z|^2)^m f(z) = 0$ for every $m > 0$. By comparison with the integral

$$\int_{\mathbf{R}^n} \frac{dx_1 \dots dx_n}{(1 + x_1^2 + \cdots x_n^2)^m} = K_n \int_0^\infty \frac{\rho^{n-1} d\rho}{(1+\rho^2)^m},$$

it is easy to see that $\sum_{z\in Z} \frac{1}{(1+|z|^2)^m} < \infty$ if $m > \frac{n}{2}$. Hence $\sum_{z\in Z} |f(z)| < \infty$. (See Problem III-3.) By Schwartz's theorem, III-4.2, $\widehat{f} \in \mathcal{S}(\widehat{E})$ and hence $\sum_{\zeta\in Z^*} |\widehat{f}(2\pi\zeta + t)| < \infty$, which guarantees the existence and continuity of $\psi(t)$. Let the subset V of $\widehat{E}$ be defined by

$$V = \{t = \sum_{i=1}^n t_i e_i^* : -\pi \le t < \pi\}.$$

Then $\{2\pi\zeta + V\}_{\zeta\in Z^*}$ form a partition of $\widehat{E}$. Hence, if $z \in Z$,

$$(i) \qquad \int_V \psi(t) e^{-i\langle z,t\rangle} \frac{dt}{(2\pi)^n} = \int_{\widehat{E}} \widehat{f}(t) e^{-i\langle z,t\rangle} \frac{dt}{(2\pi)^n} = f(z)$$

by the Fourier inversion formula.

We now define the continuous function $\psi_1(t) = \sum_{z_1\in Z} f(z_1) e^{i\langle z_1,t\rangle}$. For $z \in Z$,

$$(ii) \qquad \int_V \psi_1(t) e^{-i\langle z,t\rangle} \frac{dt}{(2\pi)^n} = f(z) \int_V \frac{dt}{(2\pi)^n} = f(z) \cdot \mathrm{Vol}(e_1^*, \dots, e_n^*).$$

(i), (ii), and the uniqueness of the Fourier coefficients imply that the two continuous functions $\psi(t)$ and $\psi_1(t) \cdot (\mathrm{Vol}(e_1^*, \dots, e_n^*))^{-1}$ are equal everywhere.

REMARKS. 1. With the above hypotheses on the choice of dt on $\widehat{E}$, it can be shown that

$$\mathrm{vol}(e_1, \ldots, e_n) \times \mathrm{vol}(e_1^*, \ldots, e_n^*) = 1.$$

Without loss of generality we may assume that $\mathrm{vol}(e_1, \ldots, e_n) = 1$. Let E be given the Euclidean structure such that $(e_1, \ldots, e_n)$ is orthonormal; then $\widehat{E}$ can be identified canonically with E, $e_j^* = e_j$, and dx and dt are identical.

2. Poisson's formula is also valid in some situations that differ slightly from that where $f \in \mathcal{S}(E)$. One of these occurs when $f \in L^1(E)$, $f \geq 0$, and $\widehat{f}$ has compact support.

3. A striking application of Poisson's formula is that if

$$g(\sigma) = \sqrt{\sigma} \sum_{n=-\infty}^{+\infty} \exp(-\sigma^2 \pi n^2),$$

then $g(\sigma) = g(\sigma^{-1})$. It suffices to take $E = \mathbf{R}$, $e_1 = 1$, and $f(x) = \mathrm{e}^{-\frac{2\pi^2 x^2}{\sigma^2}}$.

Problem III-15. *Let E be a real vector space of dimension $n > 0$, let $\widehat{E}$ be its dual, and let E be equipped with Lebesgue measure dx. It is always true that $\widehat{\widehat{E}} = E$. The canonical linear form on $E \times \widehat{E}$ is written $\langle\ ,\ \rangle$. We consider the following operators, where $a \in E$, $b \in \widehat{E}$, c (resp. d) is an invertible linear mapping from E into E (resp. from $\widehat{E}$ into $\widehat{E}$), and ${}^tc$ (resp. ${}^td$) is the transpose of c (resp. d).*

For $f \in L^2(E)$,

$$T_a f(x) = f(x-a), \quad M_b f(x) = e^{i\langle x,b\rangle} f(x), \quad H_c f(x) = f(c^{-1}x),$$

and $Uf \in L^2(\widehat{E})$ is the Fourier-Plancherel transform described in III-2.4.9.

For $g \in L^2(\widehat{E})$,

$$T_b g(t) = g(t-b), \quad M_a g(t) = e^{i\langle a,t\rangle} g(t), \quad H_d g(T) = g(d^{-1}t),$$

and $Vg \in L^2(E)$ is the Fourier-Plancherel transform.

Prove the following formulas.

(1)	$UT_a = M_a U$	(1′)	$VT_b = M_b V$
(2)	$UM_b = T_b U$	(2′)	$VM_a = T_a V$
(3)	$UH_c = \lvert det\ c\rvert H_{({}^tc)^{-1}} U$	(3′)	$VH_d = \lvert det\ d\rvert H_{({}^td)^{-1}} V$
(4)	$(H_{-\mathbf{1}_E} U)(f) = \overline{U(\overline{f})}$	(4′)	$(H_{-\mathbf{1}_{\widehat{E}}} V)(g) = \overline{V(\overline{g})}$
(5)	$U^{-1} = (2\pi)^{-n} H_{-\mathbf{1}_{\widehat{E}}} V$	(5′)	$V^{-1} = (2\pi)^{-n} H_{-\mathbf{1}_E} U$

(Here $\mathbf{1}_E$ and $\mathbf{1}_{\widehat{E}}$ are the identity operators on E and $\widehat{E}$, respectively.)

SOLUTION. Since $(\widehat{E})^\wedge = E$, it is clear that the primed formulas follow from the others. To prove (1), take $f \in L^1 \cap L^2(E)$ and verify that $UT_a f(x) = M_a U f(x)$ by writing out the Fourier integral. The fact that $L^1 \cap L^2(E)$ is dense in $L^2(E)$ implies (1); (2) is similar. (3) is proved in the same way: if $f \in L^1 \cap L^2(E)$, set $x = cx'$ in the integral

$$\int e^{i\langle x,t\rangle} f(x^{-1}x)dx = |\det c| \int e^{i\langle x', {}^t ct\rangle} f(x')dx,$$

using Theorem II-4.4.1 on the change of variable in an integral. (4) is immediate; (5) follows from the course and (4).

Problem III-16. *Use the result of Problem IV-12,*

$$\int_0^\infty x^{\alpha-1} e^{-x+ixt} \frac{dx}{\Gamma(\alpha)} = (1-it)^{-\alpha} \quad \textit{for } t \in \mathbf{R} \textit{ and } \alpha > 0,$$

with the convention for z^α with $Re(z) > 0$ made in Problem IV-12, to compute the Fourier-Plancherel transforms of the following functions in $L^2(\mathbf{R})$.
(1) $|x|^{\alpha-1} e^{-x} \mathbf{1}_{[0,+\infty)}(x)$
(2) $|x|^{\alpha-1} e^{x} \mathbf{1}_{(-\infty,0]}(x)$
(3) $|x|^{\alpha-1} e^{-x}$
(4) $-i \cdot sign(x)|x|^{\alpha-1} e^{-|x|}$
(5) $(x-a-ib)^{-n}$, with n a positive integer, a and b real, and $b \neq 0$
(6) $b(x^2+b^2)^{-1}$
(7) $x(x^2+b^2)^{-1}$
(8) $f(x)$, where $f(x)$ is a rational function with no real poles and without entire part

METHOD. For (5), use (1) and problem III-15.

SOLUTION. The first four functions are in $L^1(\mathbf{R})$ and their Fourier-Plancherel transforms are given by the integral. The transforms are
(1) $\Gamma(\alpha)(1-it)^{-\alpha}$;
(2) $\Gamma(\alpha)(1+it)^{-\alpha}$;
(3) $\Gamma(\alpha)(1+t^2)^{-\alpha}[(1+it)^\alpha + (1-it)^\alpha]$; and
(4) $\Gamma(\alpha)(1+t^2)^{-\alpha}[i(1-it)^\alpha - i(1+it)^\alpha]$. In particular, this is $2t(1+t^2)^{-1}$ if $\alpha = 1$.
The functions (5) and (8) are not necessarily in $L^1(\mathbf{R})$, and (7) is not.

We use Problem III-15, identifying $E = \widehat{E} = \mathbf{R}$. Then $U = 2\pi H_{-1} U^{-1}$. Set $f_n(x) = \frac{1}{\Gamma(n)} x^{n-1} e^{-x} \mathbf{1}_{[0,\infty)}(x)$ and $g_n(x) = (1-ix)^{-n}$. (1) implies that $Uf_n = g_n$, $U^{-1} g_n = f_n$, and hence $Ug_n = 2\pi H_{-1} f_n$. Therefore $(x-ib)^{-n} = i^n b^{-n} H_{-b} g_n(x)$ and, setting

$$f(x) = (x-a-ib)^{-n},$$

we may use Problem III-15 to write

$$f(x) = i^n b^{-n} T_a H_{-b} g_n$$

and

$$Uf = i^n b^{-n}|b| M_a H_{-1} U g_n = 2\pi i^n b^{-n} |b| M_a H_{b^{-1}} f_n.$$

Hence, for (5),

$$Uf(x) = \frac{2\pi i^n}{\Gamma(n)} \text{sign}(b) e^{iax} x^{n-1} e^{-bx} \mathbf{1}_{[0,\infty)}(bx).$$

The result of (5) can be used for (6), by writing

$$b(x^2+b^2)^{-1} = \frac{1}{2}\left[(b-ix)^{-1} + (b+ix)^{-1}\right].$$

This gives $Uf(x) = \pi e^{-|x|}$, a result already obtained in Problem III-4. For (7), if $f_b(x) = x(x^2+b^2)^{-1}$ and $g(x) = -i\text{sign}(x)e^{-|x|}$, we have seen that $U(g) = 2f_1$; hence $U(f_1) = 2\pi H_{-1} U^{-1}(f_1) = \pi H_{-1} g$. Since $f_b = b^{-1} H_b f_1$, we can use Problem III-15 to write $Uf_b = \text{sign} b \pi H_{-b^{-1}} g$, and hence

$$Uf_b(x) = -i\pi \text{ sign} x \text{ } e^{-|bx|}.$$

For (8), if Z is the set of poles of f and $z = a_z + ib_z$ for $z \in Z$ and a_z and b_z real, then $b_z \neq 0$. We know that, for every z in Z, there exists a polynomial P_z with nonzero constant term such that

$$f(x) = \sum_{z \in Z} P_z[(x - a_z - ib_z)^{-1}],$$

and the general formula for Uf follows from (5).

Problem III-17. *Compute the Fourier-Plancherel transforms of the following functions: (1)* $\mathbf{1}_{[-1,+1]}(x)$, *(2)* $\mathbf{1}_{[\alpha,\beta]}(x)$, *(3)* $\frac{\sin x}{x}$, *(4)* $\frac{\sin^2 x}{x^2}$, *and (5)* $(1-|x|)^+$.
(Here $a^+ = \max\{0, a\}$.*)*

SOLUTION. $\int_{-1}^{+1} e^{ixt} dx = 2\frac{\sin t}{t}$. Set $f(x) = \mathbf{1}_{[-1,+1]}(x)$ and $g(x) = \frac{\sin x}{x}$. Then $U(f) = 2g$.

Then, for (2), if $a = \frac{\alpha+\beta}{2}$ and $b = \frac{\beta-\alpha}{2}$, with the notation of Problem III-15,

$$(T_a H_b f)(x) = (T_a f)\left(\frac{x}{\frac{1}{2}(\beta-\alpha)}\right) = \mathbf{1}_{[\alpha,\beta]}(x).$$

Hence

$$(UT_a H_b f)(x) = b(M_a H_{b^{-1}} Uf)(x) = \frac{2}{x} e^{\frac{1}{2}(\alpha+\beta)x} \sin \frac{\beta-\alpha}{2} x.$$

For (3), $U(g) = 2\pi\overline{H^{-1}(\overline{g})} = \pi f$.

For (4), $U(f * f) = 4g^2$ since $U(f) = 2g$ and $f \in L^1(\mathbf{R})$. Hence

$$U(g^2) = 2\pi U^{-1}(g^{-2}) = \frac{\pi}{2}(f * f).$$

Thus we must compute $(f * f)(x) = \int_{-\infty}^{+\infty} f(y)f(x-y)dy$, and we easily obtain

$$(f * f)(x) = 4(1 - |x|)^+.$$

Finally, (5) follows immediately from (4) since $U(\frac{1}{4}f * f)(x) = \frac{\sin^2 x}{x^2}$.

Problem III-18. *If $f \in L^2(\mathbf{R})$ and $(U_a f)(t) = \int_{-a}^{a} e^{ixt} f(x)dx$, show that $\lim_{a\to\infty} U_a(f) = U(f)$, where U denotes the Fourier-Plancherel transform of f.*

SOLUTION. If $g \in L^\infty(\mathbf{R})$, let $(M_g f)(x) = g(x)f(x)$. It is clear that $\|M_g f\|_{L^2(\mathbf{R})} \le \|g\|_\infty \|f\|_{L^2(\mathbf{R})}$ and hence that $M_g : L^2(\mathbf{R}) \to L^2(\mathbf{R})$ is continuous. Set $g_a(x) = \mathbf{1}_{[-a,a]}(x)$. Then $U_a = UM_{g_a}$, since $L^2([-a,a]) \subset L^1([-a,a])$ implies that $g_a f \in L^1(\mathbf{R})$. Since it is clear that $M_{g_a} f \to f$ in $L^2(\mathbf{R})$ as $a \to \infty$, the result is proved.

Problem III-19. *If f and g are in $L^2(\mathbf{R})$, show that*

$$\int f(x)\widehat{g}(x)dx = \int \widehat{f}(x)g(x)dx.$$

METHOD. Use the fact that $L^1 \cap L^2(\mathbf{R})$ and $A(\mathbf{R})$ are dense in $L^2(\mathbf{R})$. (See III-2.4.7.)

SOLUTION. If $f \in L^1 \cap L^2(\mathbf{R})$ and $g \in A(\mathbf{R})$, then $\widehat{g} \in L^1 \cap L^2(\mathbf{R})$ and

$$\int f(x)\widehat{g}(x)dx = \int f(x)dx \int e^{ixt}g(t)dt = \int g(t)dt \int e^{ixt}f(x)dx$$

by Fubini. The extension to $L^2(\mathbf{R})$ is immediate.

Problem III-20. *Let $g_b(x) = i(\mathrm{sign} x)e^{-b|x|}$ for $b > 0$, let U be the Fourier-Plancherel transform in $L^2(\mathbf{R})$, and let M_{g_b} be the operator on $L^2(\mathbf{R})$ defined by $M_{g_b}f(x) = g_b(x)f(x)$. Set*

$$\mathcal{H}_b = U^{-1}M_{g_b}U.$$

(1) Show that $\mathcal{H}_b f(x) = \frac{1}{\pi}\int_{-\infty}^{+\infty} \frac{y}{y^2+b^2} f(x-y)dy$ for almost every x, if $f \in L^2(\mathbf{R})$ and $b > 0$.

METHOD. Use Problem III-16(7) to compute $U(g_b)$, then apply Problem III-19.

(2) If $f \in L^2(\mathbf{R})$, show that $\mathcal{H}_0 f = \lim_{b\downarrow 0} \frac{1}{\pi}\int_{-\infty}^{+\infty} \frac{y}{y^2+b^2} f(x-y)dy$ exists in L^2 and give its Fourier transform. Also calculate $\mathcal{H}_0^2 f$.

SOLUTION. (1) $\mathcal{H}_b f(x) = (U^{-1}M_{g_b}Uf)(x) = \frac{1}{2\pi}(H_{-1}UM_{g_b}Uf)(x) = \frac{1}{2\pi}H_{-1}U(g_b\widehat{f})(x) = \frac{1}{2\pi}\int_{-\infty}^{+\infty} \mathrm{e}^{-ixt} g_b(t)\widehat{f}(t)dt = \frac{1}{2\pi}\int_{-\infty}^{+\infty} f(t)(UM_{-a})(g_b)$ $(t)dt$ (by Problem III-19), and this is equal to

$$\frac{1}{2\pi}\int_{-\infty}^{+\infty} f(t)(T_x U)(g_b)(t)dt$$

by Problem III-15.

By Problem III-16, $U(g_b)(t) = \frac{-2t}{t^2+b^2}$. Hence

$$\mathcal{H}_b f(x) = \frac{1}{\pi}\int_{-\infty}^{+\infty} f(t)\frac{-t+x}{(t-x)^2+b^2}dt = \frac{1}{2\pi}\int_{-\infty}^{+\infty} \frac{y}{y^2+b^2} f(x-y)dy.$$

(2) It is clear that $b \mapsto M_{g_b} f$ is a continuous function from $[0,+\infty)$ to $L^2(\mathbf{R})$, and thus so is $b \mapsto \mathcal{H}_b f$. Hence

$$\mathcal{H}_0 f(x) = \lim_{b\downarrow 0}\frac{1}{\pi}\int_{-\infty}^{+\infty} \frac{y}{y^2+b^2} f(x-y)dy$$

exists in L^2, and $(U\mathcal{H}_0 f)(x) = M_{g_0}Uf(x) = i \operatorname{sign}(x)\widehat{f}(x)$. Finally, since $\mathcal{H}_0 f = U^{-1}M_{g_0}Uf$ and $g_0^2 = -1$,

$$\mathcal{H}_0^2 f = U^{-1}M_{g_0}UU^{-1}M_{g_0}Uf = -f.$$

REMARK. $\mathcal{H}_0 f$ is called the *Hilbert transform* of f.

Problem III-21. *If $f \in L^2(\mathbf{R})$ and $g \in L^1(\mathbf{R})$, show that $h(x) = \int_{-\infty}^{+\infty} f(x-y)g(y)dy$ exists for almost every x and defines a function h in $L^2(\mathbf{R})$ such that $\|h\|_{L^2} \leq \|f\|_{L^2}\|g\|_{L^1}$ and $\widehat{h} = \widehat{g}\widehat{f}$ (where $\widehat{g}$ is the Fourier transform of $g \in L^1$ and $\widehat{h}$ and $\widehat{f}$ are the Fourier-Plancherel transforms in L^2).*

METHOD. Apply Schwarz's inequality to $|f(x-y)|\,|g(y)|^{\frac{1}{2}}$ (considered as a function of y) and $|g(y)|^{\frac{1}{2}}$ and use Problem III-18.

SOLUTION.

$$H(x) = \int_{\mathbf{R}} |f(x-y)|\,|g(y)|^{\frac{1}{2}}|g(y)|^{\frac{1}{2}}dy \leq \left[\int_{\mathbf{R}} |f(x-y)|^2|g(y)|dy\right]^{\frac{1}{2}} \|g\|_{L^1}$$

by Schwarz's inequality (slightly generalized, since $H(x)$ might equal $+\infty$ for certain x). Therefore, by Fubini and the change of variables $x' = x - y$ and $y' = y$,

$$\int_{-\infty}^{+\infty} H^2(x)dx \leq \|g\|_{L^2} \int\int |f(x-y)|^2|g(y)|dxdy = \|f\|_{L^2}^2\|g\|_{L^1}^2,$$

which implies that H is finite almost everywhere, h exists almost everywhere, and $\|h\|_{L^2} \leq \|H\|_{L^2} \leq \|f\|_{L^2}\|g\|_{L^1}$.

It remains to prove that $\widehat{h} = \widehat{g}\widehat{f}$. If f_a is the restriction of f to $[-a, +a]$, we saw in Problem III-18 that $\widehat{f_a} \to \widehat{f}$ in $L^2(\mathbf{R})$ as $a \to \infty$. Since $f_a \in L^1(\mathbf{R})$, $g * f_a \in L^1$ and $\widehat{g * f_a} = \widehat{g}\widehat{f_a}$. Since $\widehat{g}$ is bounded, multiplication by $\widehat{g}$ in $L^2(\mathbf{R})$ is continuous. Hence

$$\widehat{g * f} = \lim_{a\to\infty} \widehat{g * f_a} = \lim_{a\to\infty} \widehat{g}\widehat{f_a} = \widehat{g} \lim_{a\to\infty} \widehat{f_a} = \widehat{g}\widehat{f}.$$

Problem III-22. *Let $0 < \epsilon < a$ and let $g_{\epsilon,a}(y) = (\pi y)^{-1}\mathbf{1}_{\{\epsilon\leq|y|\leq a\}}(y)$.*
(1) Compute $\lim_{\epsilon\to 0}\lim_{a\to+\infty}\widehat{g}_{\epsilon,a}(t)$, where $\widehat{g}_{\epsilon,a}$ is the Fourier transform on $L^1(\mathbf{R})$ of $g_{\epsilon,a}$. (Use Problem II-16.)
(2) For $f \in L^2(\mathbf{R})$, we set

$$\mathcal{H}_{\epsilon,a}(f) = \int_{\epsilon\leq|y-x|\leq a} \frac{f(y)dy}{x-y}.$$

(This equals $f * g_{\epsilon,a}$ in the sense of Problem III-21.) Using Problems III-20 and III-21, show that $\lim_{\epsilon\to 0}\lim_{a\to+\infty}\mathcal{H}_{\epsilon,a}(f)$ exists and coincides with the Hilbert transform of f (Problem III-20).

SOLUTION. (1)

$$\widehat{g}_{\epsilon,a}(t) = \int_{-a}^{-\epsilon} + \int_{\epsilon}^{a} \frac{e^{iyt}}{y}dy = \frac{2i}{\pi}\int_{\epsilon}^{a}\frac{\sin ty}{y}dy = \frac{2i}{\pi}\text{sign } t\int_{\epsilon|t|}^{a|t|}\frac{\sin y}{y}dy$$

By Problem II-16, $\int_0^\infty \frac{\sin x}{x}dx = \frac{\pi}{2}$. Hence

$$\lim_{\epsilon\to 0}\lim_{a\to+\infty}\widehat{g}_{\epsilon,a}(t) = i(\text{sign } t).$$

(2) Let U be the Fourier-Plancherel transform on $L^2(\mathbf{R})$. By Problem III-21, $(U\mathcal{H}_{\epsilon,a}f)(t) = \widehat{g}_{\epsilon,a}(t)(Uf)(t)$.

$$U\mathcal{H}_{0\infty}(f) = \lim_{\epsilon\to 0}\lim_{a\to 0}(U\mathcal{H}_{\epsilon,a}f)(t) = i(\text{sign } t)(Uf)(t)$$

exists in L^2; hence, by the Plancherel isomorphism, $\mathcal{H}_{0\infty}(f)$ exists and coincides with the Hilbert transform defined in Problem III-20.

Problem III-23. *A function f in $L^2(\mathbf{R})$ is called* hermitian *if $f(x) = \overline{f(-x)}$ and* skew hermitian *if $f(x) + \overline{f(-x)} = 0$. Let $\widehat{f}$ denote the Fourier-Plancherel transform of f and let $\mathcal{H}_0 f$ denote the Hilbert transform of f. (See Problems III-20 and III-22.) Prove the following statements:*

f is iff	*hermitian*	*skew-hermitian*	*real*	*purely imaginary*	*even*	*odd*
$\widehat{f}$ is iff	*real*	*purely imaginary*	*hermitian*	*skew-hermitian*	*even*	*odd*
$(\mathcal{H}_0 f)^\wedge$ is iff	*purely imaginary*	*real*	*hermitian*	*skew-hermitian*	*odd*	*even*
$(\mathcal{H}_0 f)$ is	*skew-hermitian*	*hermitian*	*real*	*purely imaginary*	*odd*	*even*

SOLUTION. The statements are easily verified if $f \in L^1 \cap L^2(\mathbf{R})$, and extend to $L^2(\mathbf{R})$ by continuity.

Problem III-24. *Compute the Hilbert transform (see Problem III-20) of each of the following functions:*

$$\begin{aligned} f_1(x) &= \tfrac{1}{\pi}\tfrac{1}{1+x^2}, & f_2(x) &= \tfrac{x}{\pi(1+x^2)}, \\ f_3(x) &= \tfrac{1}{2}\mathbf{1}_{[-1,1]}(x), & f_4(x) &= \tfrac{1}{\pi}\log\left|\tfrac{x+1}{x-1}\right|, \\ f_5(x) &= (1-|x|)^+, & f_6(x) &= \tfrac{1}{\pi}\log\left|\tfrac{x+1}{x-1}\right| + \tfrac{x}{\pi}\log\left|\tfrac{x^2-1}{x^2}\right|. \end{aligned}$$

METHOD.Use Problem III-16 for f_2 and Problem III-22 for f_6.
For f_5, $a^+ = \max\{0, a\}$.

SOLUTION. $\widehat{\mathcal{H}_0 f_1}(t) = i\ \mathrm{sign}(t)e^{-|t|} = \widehat{f_2}(t)$ by Problems III-16 and III-20. Hence $\mathcal{H}_0 f_1 = f_2$. Since $\mathcal{H}_0^2 \widehat{f} = -f$ in general, we have $\mathcal{H}_0 f_2 = -f_1$. Next,

$$\mathcal{H}_0 f_3(x) = \lim_{\epsilon \to 0} \int_{|y|\geq\epsilon,\ |x-y|\leq 1} (2\pi y)^{-1} dy = f_4(x)$$

by Problem III-22; thus $\mathcal{H}_0 f_3 = -f_4$.

Computing as in Problem III-22 shows that $\mathcal{H}_0 f_5(x) = f_6$; hence $\mathcal{H}_0 f_6 = -f_5$.

Problem III-25. *Let $\mathcal{S}$ be the vector space of C^∞ functions on $\mathbf{R}$ which, together with all their derivatives, are of rapid decrease.*
(1) Show that if $f \in \mathcal{S}$, then $\lim_{\epsilon\to 0} \int_{|x|\geq\epsilon} \frac{f(x)}{x} dx$ exists and defines a continuous linear functional (or "tempered distribution") on $\mathcal{S}$.
(2) Show that the Fourier transform of the distribution defined in (1) is the Radon measure $\mu(dt) = i\pi(\mathrm{sign}\ t)dt$.

METHOD. Split the first integral into $\{\epsilon \le |x| \le 1\} \cup \{|x| > 1\}$. Also use the fact, proved in Problem II-16, that $\int_0^\infty \frac{\sin x}{x} dx = \frac{\pi}{2}$.

SOLUTION. (1) If $f \in \mathcal{S}$, let $g_f \in \mathcal{S}$ be defined by $g_f(0) = f'(0)$ and $g_f(x) = \frac{f(x)-f(0)}{x}$ if $x \neq 0$.

$$\int_{|x|\ge\epsilon} \frac{f(x)dx}{x} = \int_{|x|\ge 1} \frac{f(x)}{x} dx + \int_{\epsilon\le|x|\le 1} \left(\frac{f(0)}{x} + g_f(x)\right) dx.$$

Since $\displaystyle\int_{\epsilon\le|x|\le 1} \frac{f(0)}{x} dx = 0$, we have

$$\lim_{\epsilon\to 0} \int_{|x|\ge\epsilon} f(x)\frac{dx}{x} = \int_{|x|\ge 1} f(x)\frac{dx}{x} + \int_{-1}^{+1} g_f(x)dx.$$

$f \mapsto \int_{|x|\ge 1} f(x)\frac{dx}{x}$ is a continuous linear functional on $\mathcal{S}$ since it is defined by the Radon measure $\mathbf{1}_{\{|x|\ge 1\}}(x)\frac{dx}{x}$. For the second term, we use the following norm on $\mathcal{S}$ (see III-4.1(ii)):

$$\|f\|_{0,1} = \sup_x(|f(x)|,\ |f'(x)|).$$

By the mean value theorem, for every x in $\mathbf{R}$ there exists θ such that $g(x) = f'(\theta x)$. Thus

$$\left|\int_{-1}^{+1} g_f(x)dx\right| \le 2\sup_x |f'(x)| \le 2\|f\|_{0,1},$$

which shows that the second term also defines a tempered distribution.
(2) Let ℓ be the distribution defined in (1) and let $\widehat{f}$ be the Fourier transform on $\mathcal{S}$ of f. By definition, $\langle \widehat{f}, \ell\rangle = \langle f, \widehat{\ell}\rangle$.

$$\langle \widehat{f}, \ell\rangle = \lim_{\epsilon\to 0} \int_{\epsilon\le|t|\le 1} \frac{dt}{t} \int_{-\infty}^{+\infty} e^{ixt} f(x)dx + \int_{|t|\ge 1} \frac{dt}{t} \int_{-\infty}^{+\infty} e^{ixt} f(x)dx.$$

The second integral exists since $\mathcal{S}$ is preserved under the Fourier transform; it is written $\lim_{T\to\infty} \int_{1\le|t|\le T}$. Then

$$\begin{aligned} \langle \widehat{f}, \ell\rangle &= \lim\nolimits_{\epsilon\to 0} \int_{-\infty}^{+\infty} f(x)dx \int_\epsilon^1 \frac{e^{ixt} - e^{-ixt}}{t} dt \\ &\quad + \lim\nolimits_{T\to\infty} \int_{-\infty}^{+\infty} f(x)dx \int_1^T \frac{e^{ixt} - e^{-ixt}}{t} dt \\ &= \int_{-\infty}^{+\infty} f(x)dx \int_0^\infty \frac{2i\sin xt}{t} = \int_{-\infty}^{+\infty} i\pi(\mathrm{sign} x) f(x)dx \end{aligned}$$

by Problem II-16.

Problem III-26. *Let $I = (a,b)$ and let $f \in L^1(I)$.*
(1) If $F(x) = \int_a^x f(t)dt$ for $x \in I$, show that $F'(x) = f(x)$ in the weak sense (III-3.3.1).
(2) If $F \in L^1(I)$ and $F' = f$ in the weak sense, show that, for $a < \alpha < \beta < b$,

$$\int_\alpha^\beta f(t)dt = F(\beta) - F(\alpha).$$

(3) Let s be a positive integer. Show that F is in H^s_{loc}, the local Sobolev space (see III-3.5.6), if and only if there exists $f \in L^2_{loc}(I)$ such that the weak derivative of order $s-1$ of F exists in the ordinary sense and satisfies

$$F^{(s-1)}(x) = F^{(s-1)}(a) + \int_\alpha^x f(t)dt$$

for all α and x in I.

SOLUTION. (1) Let $\varphi \in \mathcal{D}(I)$. Then $\varphi(b) = 0$ and, reversing the order of integration,

$$\int_a^b \left[\int_a^x f(t)dt\right] \varphi'(x)dx = \int_a^b f(t)\left[\int_t^b \varphi'(x)dx\right] dt = -\int_a^b f(t)\varphi(t)dt.$$

(2) This is exactly Lemma III-3.3.3.
(3) ($\Leftarrow$) If $f \in L^2_{loc}(I)$, then $\mathbf{1}_{(\alpha,\beta)} f \in L^1((\alpha,\beta))$ for $(\alpha,\beta) \subset I$. By (1), $F^{(s)} = f$ in the weak sense; since $F^{(k)}$ is continuous for $k < s$, $F^{(k)} \in L^2_{loc}(I)$ for $k = 0, 1, \ldots, s$.
($\Rightarrow$) Conversely, $F^{(s)} = f \in L^2_{loc}$ implies the desired formula by (2). Moreover, it follows from (1) that, since the weak derivative $F^{(s-1)}$ is continuous, it is a derivative in the ordinary sense.

Problem III-27. *Let $f \in L^2(\mathbf{R})$, with Fourier-Plancherel transform $\widehat{f}$. Prove Hermann Weyl's inequality,*

$$\left[\int_{-\infty}^{+\infty} |f(x)|^2 dx\right]^2 \leq \frac{2}{\pi}\int_{-\infty}^{+\infty} x^2|f(x)|^2 dx \times \int_{-\infty}^{+\infty} t^2|\widehat{f}(t)|^2 dt,$$

and analyze the case of equality.

METHOD. Without loss of generality, assume that f is in the Sobolev space $H^1(\mathbf{R})$. Show that $\int_{-\infty}^{+\infty} |f(x)|^2 dx = -2\mathrm{Re}\int_{-\infty}^{+\infty} x f(x)\overline{f'(x)}dx$, with the help of Problem I-15(2). Conclude by using Schwarz's inequality (Problem I-12).

SOLUTION. Without loss of generality, we assume that $\int_{-\infty}^{+\infty} x^2|f(x)|^2 dx < \infty$ and $\int_{-\infty}^{+\infty} t^2|\widehat{f}(t)|^2 dt < \infty$. This second inequality implies that $f \in$

$H^1(\mathbf{R})$ and hence that f' exists in the weak sense. Since $x|f(x)|^2$ has a derivative in the weak sense, by Problem III-26 we can perform the following integration by parts:

$$\int_{-T}^{T} x(f(x)\overline{f'(x)} + \overline{f(x)}f'(x))dx = [x|f(x)|^2]_{-T}^{+T} - \int_{-T}^{T} |f(x)|^2 dx.$$

But, since $xf(x)$ is square integrable, $T|f(T)|^2 \to 0$ as $|T| \to \infty$ by Problem I-15(2). Hence

$$\begin{aligned}\left[\int_{-\infty}^{+\infty} |f(x)|^2 dx\right]^2 &= 4\left[\mathrm{Re}\int_{-\infty}^{+\infty} xf(x)\overline{f'(x)}dx\right]^2 \\ &\leq 4\int_{-\infty}^{+\infty} x^2 f(x)^2 dx \int_{-\infty}^{+\infty} |f'(x)|^2 dx.\end{aligned}$$

Finally, we observe that $\widehat{f'}(t) = -it\widehat{f}(t)$, and hence by the Fourier-Plancherel isometry

$$\frac{1}{2\pi}\int_{-\infty}^{+\infty} t^2 f(t)^2 dt = \int_{-\infty}^{+\infty} |f'(x)|^2 dx.$$

This completes the proof of Weyl's inequality.

As for equality, we know that equality holds in the Schwarz inequality if and only if, in the present case, there exists a real λ such that $\lambda xf(x) - f'(x) = 0$ for almost every x. Now, the function $g(x) = f(x)\exp(-\frac{\lambda x^2}{2})$ admits a weak derivative in the L^1 sense locally, and this weak derivative satisfies $g'(x) = 0$ for all x. By Problem III-26, g is constant, and $f(x) = f(0)\exp(-\frac{\lambda x^2}{2})$. Such a function, if it is nonzero, can belong to $L^2(\mathbf{R})$ only if $\lambda > 0$. It is trivial to check that, conversely, $f(x) = f(0)\exp(-\frac{\lambda x^2}{2})$ does give equality in Weyl's inequality if $\lambda > 0$.

REMARK. This inequality has an interpretation in quantum mechanics, where it is known as Heisenberg's uncertainty principle. [1]

---

[1] H. Weyl, *The Theory of Groups and Quantum Mechanics* (London: Dover, 1931).

IV

Hilbert Space Methods and Limit Theorems in Probability Theory

Problem IV-1. *The points marked on the faces of two dice are, respectively,*

for the first: 1, 2, 2, 3, 3, 4*;*

for the second: 1, 3, 4, 5, 6, 8.

If X is the sum of the points obtained by throwing the two dice, compute $P[X = k]$ for integer k. Answer the same question for ordinary dice.

SOLUTION. One solution consists of trying to reduce this to a uniform probability problem. Distinguish the faces of the dice by assigning the face of each die a letter in the set $A = \{a, b, c, d, e, f\}$. Consider the following two functions X_1 and X_2 on A:

$$X_1(a) = 1, \quad X_1(b) = X_1(c) = 2, \quad X_1(D) = X_1(e) = 3, \quad X_1(f) = 4,$$

$$X_2(a) = 1, \quad X_2(b) = X_2(c) = 4, \quad X_2(d) = 5, \quad X_2(e) = 6, \quad X_2(f) = 8.$$

The probability space Ω is then $A \times A$ with the uniform probability $P[\{(\omega_1, \omega_2)\}] = \frac{1}{36}$. On Ω, define the random variable $X(\omega_1, \omega_2) = X_1(\omega_1) + X_2(\omega_2)$, for which the table below gives the values for every pair $\omega = (\omega_1, \omega_2)$.

	1	a	b	c	d	e	f
2		1	2	2	3	3	4
a	1	2	3	3	4	4	5
b	3	4	5	5	6	6	7
c	4	5	6	6	7	7	8
d	5	6	7	7	8	8	9
e	6	7	8	8	9	9	10
f	8	9	10	10	11	11	12

The distribution of X can be determined by counting:

k	2	3	4	5	6	7	8	9	10	11	12
$36P[X = k]$	1	2	3	4	5	6	5	4	3	2	1

For a second solution, consider the independent random variables X_1 and X_2 representing the points which appear on each die. Instead of using the characteristic functions to compute the distribution of the sum $X = X_1 + X_2$, use the generating functions:

$$\mathbf{E}(s^{X_1}) = \frac{1}{6}(s + 2s^2 + 2s^3 + s^4), \quad \mathbf{E}(s^{X_2}) = \frac{1}{6}(s + s^3 + s^4 + s^5 + s^6 + s^8),$$

and

$$\begin{aligned}\mathbf{E}(s^X) &= \mathbf{E}(s^{X_1+X_2}) = \mathbf{E}(s^{X_1})\mathbf{E}(s^{X_2}) \\ &= \tfrac{1}{36}(s^2 + 2s^3 + 3s^4 + 4s^5 + 5s^6 + 6s^7 + 5s^8 + 4s^9 + 3s^{10} + 2s^{11} + s^{12}).\end{aligned}$$

If we consider the second question in the same way, we find that the sum of the points for ordinary dice is the *same* as that above.

Problem IV-2. *The random variable X is called a geometric distribution with parameter p, $0 < p < 1$, if*

$$P[X = k] = (1 - p)^{k-1}p, \quad k = 1,\ 2,\ 3, \ldots$$

Compute $\mathbf{E}(X)$ by using Problem I-6(1).

SOLUTION. $P[X > n] = p \sum_{k=n+1}^{\infty} (1 - p)^{k-1} = (1 - p)^n$. Hence

$$\mathbf{E}[X] = \sum_{n=0}^{\infty} P[X > n] = \frac{1}{p}.$$

Problem IV-3. *Suppose that δ_a is the Dirac measure at a, $p \in (0,1)$, and $\lambda > 0$. Consider the following two probability measures on $\mathbf{N}$:*

$$\nu_p = (1-p)\delta_0 + p\delta_1 \quad \textit{(Bernouilli distribution with parameter } p\textit{)}$$

$$\mu_\lambda = \sum_{k=0}^{\infty} e^{-\lambda} \frac{\lambda^k}{k!} \delta_k \quad \textit{(Poisson distribution with parameter } \lambda\textit{)}$$

(1) Show that the vague limit of the sequence $\{\nu_{\frac{\lambda}{n}}^{*n}\}_{n>\lambda}$ is μ_λ and that $\mu_{\lambda_1} * \mu_{\lambda_2} = \mu_{\lambda_1+\lambda_2}$.
(2) Let $0 < p < 1$. Consider the measure m_p on $\mathbf{N}^2$ concentrated at the points $(0,0)$, $(0,1)$, $(1,1)$, and $(k,0)$ with $k \geq 2$ (note the absence of $(1,0)$), such that X has distribution μ_p and Y has distribution ν_p if (X,Y) has distribution m_p. Compute m_p and conclude that $P(X \neq Y) \leq 2p^2$. (Use the fact that $e^{-p} \geq 1-p$.)
(3) If (X,Y) is an arbitrary variable in $\mathbf{N}^2$ and $A \subset \mathbf{N}$, show that

$$|P(X \in A) - P(Y \in A)| \leq P(X \neq Y).$$

(4) Let $(X_1,Y_1), \dots, (X_n,Y_n)$ be n independent random variables with values in $\mathbf{N}^2$ and with distributions $m_{p_1}, m_{p_2}, \dots, m_{p_n}$. Let $A \subset \mathbf{N}$. Use (2) and (3) to show that

$$|P(X_1 + \cdots X_n \in A) - P(Y_1 + \cdots + Y_n \in A)| \leq 2\sum_{j=1}^{n} p_j^2$$

by using (2) and (3).
(5) If $n > \lambda$ and $A \subset \mathbf{N}$, show that

$$|\nu_{\frac{\lambda}{n}}^{*n}(A) - \mu_\lambda(A)| \leq \frac{2\lambda^2}{n}.$$

SOLUTION. (1) *First solution.* Let $p \in (0,1)$ and let $q = 1-p$. Then

$$\nu_p^{*n} = (q\delta_0 + p\delta_1)^{*n} = \sum_{k=0}^{n} C_n^k p^k q^{n-k} \delta_k. \quad \text{(Binomial distribution)}$$

Hence

$$\nu_{\frac{\lambda}{n}}^{n}(k) = \frac{n(n-1)\dots(n-k+1)}{k!} \frac{\lambda^k}{n^k} (1 - \frac{\lambda}{n})^{n-k} \to \mu_\lambda(k) \quad \text{as } n \to \infty,$$

which implies that $\nu_{\frac{\lambda}{n}}^{*n}$ converges vaguely to μ_λ.
Second solution. We can use the Fourier transform of $\nu_{\frac{\lambda}{n}}^{*n}$:

$$\widehat{\nu_{\frac{\lambda}{n}}^{*n}}(t) = (1 - \frac{\lambda}{n} + \frac{\lambda}{n} e^{it})^n \to \exp(\lambda(e^{it} - 1)) = \widehat{\mu}_\lambda.$$

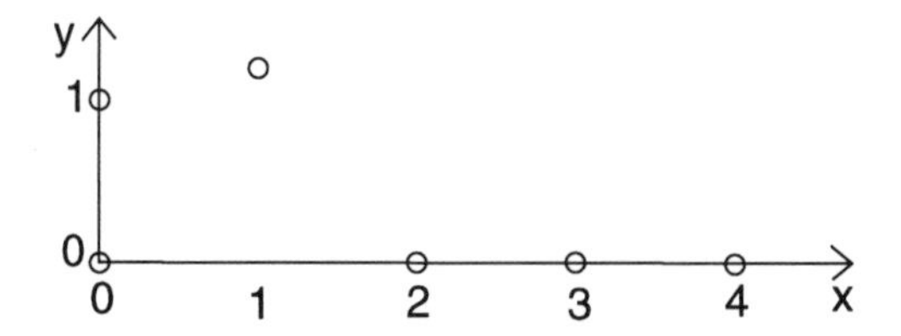

To see that $\mu_{\lambda_1} * \mu_{\lambda_2} = \mu_{\lambda_1+\lambda_2}$, we can either compute

$$(\mu_{\lambda_1} * \mu_{\lambda_2})(n) = \sum_{k=0}^{n} \mathrm{e}^{-\lambda_1}\frac{\lambda_1^k}{k!}\mathrm{e}^{-\lambda_2}\frac{\lambda_2^{n-k}}{(n-k)!} = \mu_{\lambda_1+\lambda_2}(n)$$

by the binomial formula, or use the Fourier transform.

Since $P[X=1] = p\mathrm{e}^{-p}$, it is clear that $m_p((1,1)) = p\mathrm{e}^{-p}$; since

$$P[Y=1] = p = m_p((0,1)) + m_p(1,1),$$

$m_p(0,1) = p(1-\mathrm{e}^{-p})$; and since

$$P[X=0] = \mathrm{e}^{-p} = m_p((0,0)) + m_p((0,1)),$$

$m_p((0,0)) = \mathrm{e}^{-p} - p + p\mathrm{e}^{-p}$. Moreover, it is clear that $m((k,0)) = \mathrm{e}^{-p}\frac{p^k}{k!}$. Hence

$$P(X=Y) = \mathrm{e}^{-p} - p + 2p(1-p) = 1-2p^2.$$

(3)

$$\begin{aligned}\{X \in A\} &= \{X = Y \in A\} \cup \{X \neq Y \text{ and } X \in A\} \\ &\subset \{Y \in A\} \cup \{X \neq Y\}\end{aligned}$$

It follows that $P(X \in A) \leq P(Y \in A) + P(X \neq Y)$, and by symmetry $P(Y \in A) \leq P(X \in A) + P(X \neq Y)$.
(4) Apply (3) to $X = X_1 + \cdots + X_n$ and $Y = Y_1 + \cdots + Y_n$ and note that $\{X \neq Y\} \subset \cup_{j=1}^n \{X_j \neq Y_j\}$. Hence

$$|P[X \in A] - P[Y \in A]| \leq P[X \neq Y] \leq \sum_{j=1}^{n} P[X_j \neq Y_j] \leq \sum_{j=1}^{n} 2p_j^2.$$

(5) Apply (4) to the case $p_j = \frac{\lambda}{n}$ for $j = 1, \ldots, n$. Note that the distribution of $Y_1 + \cdots + Y_n$ is $\nu^*_{\frac{\lambda}{n}}$ and that of $X_1 + \cdots + X_n$ is μ_λ by (1).

REMARKS. The approximation of the binomial distribution by the Poisson distribution is both elementary and essential for applications. (5) gives an upper bound for the error committed by replacing a binomial distribution ν_p^n by a Poisson distribution μ_{np}, and (4) treats the case of experiments that are independent but not identical. The method above is due to J.L. Hodges and L. Le Cam (1960).

Problem IV-4. *On a probability space $(\Omega, \mathcal{A}, P)$, we define a random variable N with positive integer values and random variables $\{X_n\}_{n\geq 1}$, with values in a measurable space $(I, \mathcal{B})$, such that the X_n all have the same distribution m but are not necessarily independent.*
(1) Show that the distribution μ of X_N is absolutely continuous with respect to m.
(2) If $f(x) = \frac{d\mu}{dm}(x)$ and $\alpha > 0$, show that

$$\mathbf{E}(N^\alpha) \geq \frac{1}{1+\alpha}\int_I f^{\alpha+1}(x)dm(x).$$

METHOD. If $B(y) = \{x \in I : f(x) > y\}$, show that

$$\mu(B(y)) \leq \sum_{n\leq y} P[X_n \in B(y)] + P[N > y] \tag{i}$$

and use Problem I-6.

(3) Show that $1 + \mathbf{E}(\log N) \geq \int f(x)\log f(x)dm(x)$ by letting $\alpha \downarrow 0$ in (2) and using the monotone convergence theorem.

SOLUTION. (1) If $B \in \mathcal{B}$, then

$$\mu(B) = P[X_N \in B] = \sum_{n=1}^{\infty} P[X_n \in B \ \text{ and } \ N = n]. \tag{ii}$$

Consequently $m(B) = 0$ implies $\mu(B) = 0$, since

$$0 = m(B) \geq P[X_n \in B \ \text{ and } \ N = n] \quad \text{for all } n.$$

(2) Inequality (i) clearly follows from (ii), and implies

$$\mu(B(y)) \leq y\, m(B(y)) + P[N > y].$$

Since

$$\alpha y^{\alpha-1}\mu(B(y)) \leq \frac{\alpha}{\alpha+1}(\alpha+1)y^\alpha m(B(y)) + \alpha y^{\alpha-1}P[N > y],$$

it follows from Problem I-6 that

$$\int_I f^\alpha(x)d\mu(x) \leq \frac{\alpha}{\alpha+1}\int_I f^{\alpha+1}(x)dm(x) + \int_\Omega N^\alpha(\omega)dP(\omega).$$

Since $\int_I f^\alpha d\mu = \int_I f^{\alpha+1}dm$, this gives the desired inequality.

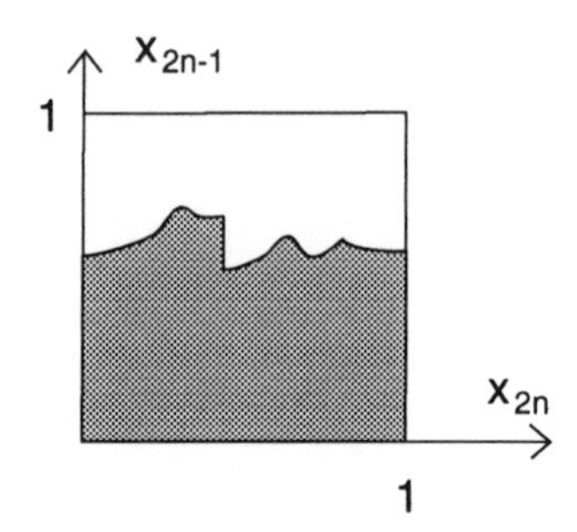

(3) For $a > 0$, the derivative of the function $f(\alpha) = \frac{a^\alpha - 1}{\alpha}$ has the same sign as $\alpha \log a - 1 + a^{-\alpha} \geq 0$, and $f(\alpha)$ decreases to $\log a$ as $\alpha \downarrow 0$.

Since $\int f(x)dm(x) = 1$, (2) implies that

$$\mathbf{E}\left[\frac{N^\alpha - 1}{\alpha}\right] \geq \frac{1}{1+\alpha}\left[\int_I \frac{f^\alpha - 1}{\alpha} f dm - 1\right].$$

The desired inequality follows from the monotone convergence theorem.

Problem IV-5. *With the notation of Problem IV-4, we take $I = [0,1]$, $\mathcal{B} =$ the Borel algebra, and $m =$ Lebesgue measure, and we assume that the $\{X_n\}_{n\geq 1}$ are independent. Let $f : I \to [0, +\infty)$ be a nonnegative measurable function, bounded by a number $b > 1$, which satisfies $\int_0^1 f(x)dx = 1$. Let*

$$N = \inf\{2n : bX_{2n-1} \leq f(X_{2n})\}.$$

Show that X_N has density $\frac{d\mu}{dm} = f$.

SOLUTION. Since the graph of the function $x \mapsto \frac{1}{b} f(x)$ lies in the square $[0,1]^2$ (see the figure), it is clear that for $B \in \mathcal{B}$

$$P[X_{2n} \in B \text{ and } X_{2n-1} \leq \frac{1}{b} f(X_{2n})] = \int_B \frac{f(x)}{b} dx.$$

Moreover, $P[N \geq 2n] = (1 - \frac{1}{b})^{n-1}$ and

$$\begin{aligned} &P[X_{2n} \in B \text{ and } N = 2n] \\ &\quad = P[X_{2n} \in B \text{ and } X_{2n-1} \leq \tfrac{1}{b} f(X_{2n}) \text{ and } N \geq 2n] \\ &\quad = (1 - \tfrac{1}{b})^{n-1} \textstyle\int_B \frac{f(x)}{b} dx. \end{aligned}$$

(The last equality is justified by the fact that the event $N \geq 2n$, depending only on $X_1, X_2, \ldots, X_{2n-2}$, is independent of (X_{2n-1}, X_{2n}).) Then

$$\mu(B) = \sum_{n=1}^{\infty} (1 - \frac{1}{b})^{n-1} \frac{1}{b} \int_B f(x)dx = \int_B f(x)dx.$$

REMARK. This procedure for constructing a random variable of given density f on $[0,1]$ from uniform random variables was invented by J. Von Neumann in 1951.

Problem IV-6. *(1) Let Y be a positive random variable. Show that, for all $y > 0$,*

$$P(Y \geq y) \leq \frac{1}{y}\mathbf{E}(Y) \quad \textit{(Chebyshev's inequality).}$$

(2) Let X be a real random variable such that $\mathbf{E}(X^2) < \infty$. If $m = \mathbf{E}(X)$, show that, for all $t > 0$,

$$P(|X - m| \geq t) \leq \frac{1}{t^2}\mathbf{E}((X - m)^2) \quad \textit{(Bienaimé's inequality).}$$

(3) Let $\{X_n\}_{n=1}^{\infty}$ be a sequence of independent real random variables with the same distribution and such that $\mathbf{E}(X_1^2) < \infty$. If $m = \mathbf{E}(X_1)$, show that, for all $\epsilon > 0$ and for all $\alpha \in [0, \frac{1}{2})$,

$$P\left[\left|\frac{X_1 + \cdots + X_n}{n} - m\right| \geq \frac{\epsilon}{n^\alpha}\right] \to 0 \textit{ as } n \to 0$$

(weak law of large numbers).

(4) Let $\{X_n\}_{n=1}^{\infty}$ be a sequence of independent real random variables with the same distribution, for which there exists $k > 0$ such that $\mathbf{E}[\exp k|X_1|] < \infty$. If $m = \mathbf{E}(X_1)$, show that for every $\epsilon > 0$ there exists q in $(0,1)$ such that

$$P\left[\left|\frac{X_1 + \cdots + X_n}{n} - m\right| \geq \epsilon\right] \leq 2q^n.$$

Conclude that $\frac{1}{n}(X_1 + \cdots + X_n) \to m$ almost surely as $n \to \infty$ (strong law of large numbers).

METHOD. Show that $m = \frac{d}{ds}[\mathbf{E}(\exp(sX_1))]_{s=0}$ and apply Chebyshev's inequality to $Y = \exp(s(X_1 + \cdots X_n))$.

SOLUTION. (1) $Y = Y\mathbf{1}_{\{Y<y\}} + Y\mathbf{1}_{\{Y \geq y\}} \geq y\mathbf{1}_{\{Y \geq y\}}$. Hence

$$\mathbf{E}(Y) \geq y\mathbf{E}(\mathbf{1}_{\{Y \geq y\}}) = yP[Y \geq y].$$

(2) Apply (1) to $Y = (X - m)^2$ and $y = t^2$.
(3) Apply (2) to $X = \frac{1}{n}(X_1 + \cdots + X_n)$. Then $\mathbf{E}(X) = m$ and $\mathbf{E}((X - m)^2) = \frac{n}{n^2}\mathbf{E}((X_1 - m)^2)$. Thus

$$P\left[\left|\frac{1}{n}(X_1 + \cdots + X_n) - m\right| \geq \frac{\epsilon}{n^\alpha}\right] \leq \frac{n^{2\alpha}}{n\epsilon^2}\mathbf{E}((X_1 - m)^2) \to 0 \quad \text{as } n \to \infty.$$

(4) If $|s| \le \frac{\alpha}{2}$, then $\exp(sX_1) \le \exp(\frac{\alpha}{2}|X_1|)$. By the mean value theorem, $|\frac{1}{s}[\exp(sX_1) - 1]| \le |X_1| \exp(\frac{|\alpha|}{2}|X_1|)$. Since $|x| \exp(\frac{\alpha}{2}|x|) < \exp(\alpha|x|)$ for x sufficiently large, $|X_1| \exp(\frac{\alpha}{2}|X_1|)$ is integrable. We can therefore use the dominated convergence theorem to assert that

$$\lim_{s\to 0} \mathbf{E}[\frac{1}{s}(\exp(sX_1 - 1))] = \mathbf{E}[\lim_{s\to 0} \frac{1}{s}(\exp(sX_1 - 1))] = m.$$

Next, set

$$\begin{aligned} A_n &= \{\tfrac{1}{n}(X_1 + \cdots + X_n) - m \ge \epsilon\} \\ B_n &= \{\tfrac{1}{n}(X_1 + \cdots + X_n) - m \le \epsilon\} \end{aligned}$$

Applying (1) shows that, for $\theta \le s < \alpha$,

$$\begin{aligned} P[A_n] &= P[\exp(s(X_1 + \cdots + X_n)) \ge \exp(ns(\epsilon + m))] \\ &\le e^{-ns(\epsilon+m)} \mathbf{E}[\exp(s(X_1 + \cdots + X_n))]. \end{aligned}$$

Set $\varphi(s) = e^{-s(\epsilon+m)} \mathbf{E}(\exp(sX_1))$. Then $P(A_n) \le (\varphi(s))^n$. To see that there exists $s \in (0, \alpha]$ such that $\varphi(s) < 1$, note that $\varphi'(0) = -\epsilon < 0$ and that $\varphi(0) = 1$. Hence there exists $s_1 \in (0, \alpha]$ such that $q_1 = \varphi(s_1) \in (0, 1)$.

Applying this result to $(-X_n)_{n=1}^{\infty}$ shows that there exists q_2 in $(0, 1)$ such that $P(B_n) \le q_2^n$. Taking $q = \max\{q_1, q_2\}$ gives the desired result.

We conclude that $\sum_{n=1}^{\infty} P[|\frac{1}{n}(X_1 + \cdots + X_n) - m| \ge \epsilon] < \infty$ for all $\epsilon > 0$. By the Borel-Cantelli lemma, this implies that

$$\limsup_{n\to\infty} |\frac{1}{n}(X_1 + \cdots + X_n) - m| \le \epsilon \quad \text{almost surely,}$$

and hence that

$$\limsup_{n\to\infty} \frac{1}{n}(X_1 + \cdots + X_n) = m \quad \text{almost surely.}$$

REMARKS. (3) and (4) are brief proofs of the laws of large numbers, which are valid under weaker hypotheses: existence of the derivative at the origin, im, of $\mathbf{E}(\exp(itX_1))$ for the weak law and existence of $\mathbf{E}(X_1) = m$ for the strong law. (See Problem IV-7.) These weaker hypotheses are, in a sense, necessary and sufficient.

Problem IV-7. *Let $\{X_n\}_{n=1}^{\infty}$ be a sequence of nonnegative real random variables with the same distribution, such that X_j and X_n are independent for every pair (j, n) with $j \ne n$. Assume that $\mathbf{E}(X_1) < \infty$. Set $S_n = \sum_{j=1}^{n} X_j$, $Y_n = X_n \mathbf{1}_{\{X_n \le n\}}$, and $S_n^* = \sum_{j=1}^{n} Y_j$. The goal of this problem is to prove the law of large numbers:*

$$P\left[\lim_{n\to\infty} \frac{S_n}{n} = \mathbf{E}(X_1)\right] = 1.$$

(1) Using Problem I-6, show that $\mathbf{E}(X_1) < \infty$ *implies* $\sum_{n=1}^{\infty} P[X_n \neq Y_n] < \infty$. *Using the Borel-Cantelli lemma (I-5.2.8), conclude that* $\lim_{n\to\infty}(S_n - S_n^*)$ *exists with probability* 1.
(2) Show that $\lim_{n\to\infty} \frac{1}{n}\mathbf{E}(S_n^*) = \mathbf{E}(X_1)$.
(3) Let α *be a real number greater than* 1 *and let* k_n *be the integer part of* α^n. *Prove the existence of a constant* C_1 *such that* $\sum\{k_n^{-2} : n \text{ such that } k_n \geq j\} \leq C_1 j^{-2}$. *With the help of Bienaimé's inequality, conclude that*

$$\sum_{n=1}^{\infty} P\left\{\left|\frac{S_{k_n}^* - \mathbf{E}(S_{k_n}^*)}{k_n}\right| \geq \epsilon\right\} \leq \frac{C_1}{\epsilon^2}\sum_{j=1}^{\infty}\frac{1}{j^2}\mathbf{E}(Y_j^2).$$

Then prove that $\sum_{j=1}^{\infty}\frac{1}{j^2}\mathbf{E}(Y_j^2) < \infty$.
(4) Deduce from (2), (3), and the Borel-Cantelli lemma that

$$P\left[\lim_{n\to\infty}\frac{S_{k_n}^*}{k_n} = \mathbf{E}(X_1)\right] = 1,$$

then from (1) that

$$P\left[\lim_{n\to\infty}\frac{S_{k_n}}{k_n} = \mathbf{E}(X_1)\right] = 1.$$

(5) Prove that, for every $\alpha > 1$,

$$P\left[\alpha^{-1}\mathbf{E}(X_1) \leq \liminf_{n\to\infty}\frac{S_n}{n} \leq \limsup_{n\to\infty}\frac{S_n}{n} \leq \alpha\mathbf{E}(X_1)\right] = 1.$$

Deduce the law of large numbers from this.

SOLUTION.
(1)

$$\begin{aligned}\sum_{n=1}^{\infty} P[X_n \neq Y_n] &= \sum_{n=1}^{\infty} P[X_n > n] \\ &= \sum_{n=1}^{\infty} P[X_1 > n] \\ &\leq \int_0^{\infty} P[X_1 > x]dx = \mathbf{E}(X_1) < \infty.\end{aligned}$$

Hence, with probability 1, there exists an integer N such that $X_n = Y_n$ if $n > N$. Hence $\lim_{n\to\infty} S_n - S_n^* = \sum_{j=1}^{N}(X_j - Y_j)$.
(2) $\frac{1}{n}\mathbf{E}(S_n^*) = \frac{1}{n}\sum_{j=1}^{N}\mathbf{E}(Y_j)$. But $X_n - Y_n \downarrow 0$; by monotone convergence, $\mathbf{E}(Y_n) \to \mathbf{E}(X_1)$ and $\frac{1}{n}\sum_{j=1}^{n}\mathbf{E}(Y_j) \to \mathbf{E}(X_1)$ as $n \to \infty$.
(3) Since $\frac{1}{\alpha^n} \to 0$ as $n \to \infty$, there exists an integer N such that $n \geq N$ implies $\frac{1}{\alpha^{n-1}} \leq \alpha - 1$; hence $k_{n-1} \leq \alpha^{n-1} \leq \alpha^n - 1 < k_n \leq \alpha^n$. This

implies that, for $j \geq J \geq \alpha^{N-1}$,

$$\begin{aligned}
&\sum\{k_n^{-2} : n \text{ such that } k_n \geq j\} \\
&\leq \sum\{\alpha^{-2(n-1)} : n \text{ such that } \alpha^{n-1} \geq j\} \\
&= \frac{1}{1-\alpha^{-2}} \frac{1}{\alpha^{2(N_j-1)}} \quad \text{(where } N_j \text{ is the first integer such that } \alpha^{N_j-1} \geq j) \\
&\leq \frac{1}{1-\alpha^{-2}} \frac{1}{j^2}.
\end{aligned}$$

Take C large enough that $\sum\{k_n^{-2} : n \text{ such that } k_n \geq j\} \leq \frac{C}{j^2}$ and let $C_1 = \max\{C, (1-\alpha^{-2})^{-1}\}$.

Next, set $A_n(\epsilon) = \{|S^*_{k_n} - \mathbf{E}(S^*_{k_n})| \geq \epsilon k_n\}$. By Bienaimé's inequality,

$$P[A_n(\epsilon)] \leq \frac{1}{\epsilon^2 k_n^2} \sigma^2(S^*_{k_n}),$$

where the variance $\sigma^2(S^*_{k_n})$ equals $\sum_{j=1}^{k_n} \sigma^2(Y_j)$ because the $(X_j)_{j=1}^\infty$ are pairwise independent. Noting that $\sigma^2(Y_j) = \mathbf{E}(Y_j^2) - (\mathbf{E}(Y_j))^2 \leq \mathbf{E}(Y_j^2)$, we can write

$$\sum_{n=1}^\infty P[A_n(\epsilon)] \leq \frac{1}{\epsilon^2} \sum_{n=1}^\infty \frac{1}{k_n^2} \sum_{j=1}^{k_n} \mathbf{E}(Y_j^2) \leq \frac{C_1}{\epsilon^2} \sum_{j=1}^\infty \frac{1}{j^2} \mathbf{E}(Y_j^2).$$

Now consider the last sum. Let μ be the distribution of X; then

$$\begin{aligned}
\sum_{j=1}^\infty \frac{1}{j^2} \mathbf{E}(Y_j^2) &= \sum_{j=1}^\infty \frac{1}{j^2} \int_0^j x^2 \mu(dx) \\
&= \sum_{j=1}^\infty \frac{1}{j^2} \sum_{k=0}^{j-1} \int_k^{k+1} x^2 \mu(dx) \\
&= \sum_{k=1}^\infty \int_k^{k+1} x^2 \mu(dx) \sum_{j=k+1}^\infty \frac{1}{j^2}.
\end{aligned}$$

Since $\sum_{j=k+1}^\infty \frac{1}{j^2} \leq \int_{k+1}^\infty \frac{dx}{x^2} = \frac{1}{k+1}$ and $\frac{x^2}{k+1} \leq x$ if $k \leq x \leq k+1$, we conclude that

$$\sum_{j=1}^\infty \frac{1}{j^2} \mathbf{E}(Y_j^2) \leq \sum_{k=0}^\infty \int_k^{k+1} x\mu(dx) = \mathbf{E}(X_1) < \infty.$$

(4) By the Borel-Cantelli lemma, if $B(\epsilon)^c = \cap_{n=1}^\infty \cup_{k=n}^\infty A_k(\epsilon)$, then $P(B(\epsilon)) = 1$ and hence $P(\cap_k B(\frac{1}{k})) = 1$. Since

$$B(\epsilon) = \left\{ \limsup_{n\to\infty} \left| \frac{S^*_{k_n} - \mathbf{E}(S^*_{k_n})}{k_n} \right| \leq \epsilon \right\},$$

it follows from (1) and (2) that $P[\lim_{n\to\infty}\frac{S_{k_n}}{k_n}=\mathbf{E}(X_1)]=1$.
(5) If $k_{n-1}\le j\le k_n$,

$$\frac{k_{n-1}}{j}\frac{S_{k_{n-1}}}{k_{n-1}}\le\frac{S_j}{j}\le\frac{k_n}{j}\frac{S_{k_n}}{k_n}\le\alpha\frac{S_{k_n}}{k_n}.$$

Since $\frac{k_{n-1}}{j}\ge\frac{k_{n-1}}{k_n}\to\alpha^{-1}$ as $n\to\infty$, we have $P[C(\alpha)]=1$, with

$$C(\alpha)=\left\{\alpha^{-1}\mathbf{E}(X_1)\le\liminf_{j\to+\infty}\frac{S_j}{j}\le\limsup_{j\to\infty}\frac{S_j}{j}\le\alpha\mathbf{E}(X_1)\right\}.$$

Hence $P[\cap_{k=1}^{\infty}C(1+\frac{1}{k})]=1$; this is the law of large numbers.

REMARKS. 1. The restriction $X_1\ge0$ (essential for (5)) is hardly significant; this proof can be extended to nonpositive random variables by writing $X_n=X_n^+-X_n^-$, where $X_n^+=\max\{0,X_n\}$.
2. This elementary proof is due to N. Etemadi (1981).

Problem IV-8. *Let $\{X_m\}_{n=1}^{\infty}$ be independent real random variables with the same distribution and such that $\mathbf{E}(X_1)=0$ and $0<\mathbf{E}(X_1^2)<\infty$. Let $S_n=X_1+\cdots X_n$.*
(1) Show that $\lim_{n\to\infty}P(S_n\ge0)=\frac{1}{2}$ by using Laplace's theorem (IV-4.3.1) and Problem II-22.
(2) Use the preceding result and the weak law of large numbers proved in Problem IV-6(3) (that $\lim_{n\to\infty}P[|\frac{S_n}{n^{\frac{3}{4}}}|\ge\epsilon]=0$ for all $\epsilon>0$) to show that $\lim_{n\to\infty}[\mathbf{E}(\exp(-S_n)\mathbf{1}_{S_n\ge0})]^{\frac{1}{n}}=1$.

SOLUTION. (1) If $\sigma^2=\mathbf{E}(X_1^2)$, Laplace's theorem states that the distribution μ_n of $\frac{S_n}{\sigma\sqrt{n}}$ converges narrowly to $\nu(dx)=\exp(-\frac{x^2}{2})\frac{dx}{\sqrt{2\pi}}$. By Problem II-2(3), $\mu_n([0,+\infty))\to\nu([0,+\infty))=\frac{1}{2}$.
(2)

$$\begin{aligned}&[\mathbf{E}(\exp(-S_n)\mathbf{1}_{\{S_n\ge0\}})]^{n^{-\frac{3}{4}}}\\ &\quad\ge\mathbf{E}[\exp(-n^{-\frac{3}{4}}S_n)\mathbf{1}_{\{S_n\ge0\}}]\\ &\quad\ge(1-\epsilon)P[\{S_n\ge0\}\cap\{\exp(-n^{-\frac{3}{4}}S_n)\ge1-\epsilon\}]\\ &\quad\ge(1-\epsilon)\left(P[S_n\ge0]-P[\exp(-n^{\frac{3}{4}}S_n)<1-\epsilon]\right).\end{aligned}$$

Hence $\liminf_{n\to\infty}[\mathbf{E}(\exp(-S_n)\mathbf{1}_{\{S_n\ge0\}})]^{n^{-\frac{3}{4}}}\ge(1-\epsilon)\frac{1}{2}>0$, and therefore $\liminf_{n\to\infty}[\mathbf{E}(\exp(-S_n)\mathbf{1}_{\{S_n\ge0\}})]^{\frac{1}{n}}\ge1$. Since $\exp(-S_n)\mathbf{1}_{\{S_n\ge0\}}\le1$, the result is proved.

Problem IV-9. *(1) If X and Y are independent real random variables, show that $P(X+Y\ge a+b)\ge P(X\ge a)P(Y\ge b)$ for all real a and b.*

(2) Let $\{X_n\}_{n=1}^{\infty}$ be a sequence of real independent random variables with the same distribution, and set $S_0 = 0$ and $S_n = X_1 + \cdots + X_n$. Let s be a fixed real number. Set $p_n = P[S_n \geq ns]$. Show that $p_{n+m} \geq p_n p_m$ for all m, $n \geq 0$ and that, for $n > 0$, $p_n = 0$ if and only if $p_1 = 0$.
(3) If the sequence $\{a_n\}_{n=0}^{\infty}$ of nonnegative real numbers is such that $a_{n+m} \leq a_n + a_m$ for all m, $n \geq 0$, show that $\lim_{n\to\infty} \frac{a_n}{n} = \inf_{d>0} \frac{a_d}{d}$. Conclude that $\lim_{n\to\infty} \sqrt[n]{p_n} = \alpha(s)$ exists.

SOLUTION. (1) Immediate, since $\{(x,y) : x + y \geq a + b\} \supset \{(x,y) : x \geq a,\ y \geq b\}$.
(2) If n and m are nonnegative, apply (1) to $X = S_n$, $Y = S_{n+m} - S_n$, $a = ns$, and $b = ms$. Since $P[Y \geq b] = p_m$, we have $p_{n+m} \geq p_n p_m$. It follows that $p_n \geq (p_1)^n$ if $n > 0$; hence $p_1 = 0$ if $p_n = 0$. Conversely, if $p_1 = 0$, then $P[s - X_k > 0] = 1$ for $k = 1, 2, \ldots, n$. Reasoning as in (1),

$$1 = P[\bigcap_{k=1}^{n} \{s - X_k > 0\}] \leq P[\sum_{k=1}^{n} s - X_k > 0] = 1 - p_n$$

(3) If d is a positive integer, let q_n and r_n be nonnegative integers such that $n = q_n d + r_n$ and $r_n < d$. Then $\frac{a_n}{n} \leq \frac{q_n a_d + a_{r_n}}{q_n d + r_n}$, and hence $\limsup_{n\to\infty} \frac{a_n}{n} \leq \frac{a_d}{d}$ for all $d > 0$. Thus

$$0 \leq \liminf_{n\to\infty} \frac{a_n}{n} \leq \limsup_{n\to\infty} \frac{a_n}{n} \leq \inf_{d>0} \frac{a_d}{d}.$$

It follows that $\lim_{n\to\infty} \frac{a_n}{n}$ exists and equals $\inf_{d>0} \frac{a_d}{d}$.
If $p_1 = 0$, (2) shows that $p_n = 0$ for every n. If $p_1 > 0$, set $a_n = -\log p_n$ and $a_0 = 0$. Then (2) implies that $0 \leq a_n < \infty$ and $a_{n+m} \leq a_n + a_m$ for all n, $m \geq 0$. Hence $\lim_{n\to\infty} \sqrt[n]{p_n}$ exists and equals $\sup_{d>0} \sqrt[d]{p_d}$.

Problem IV-10. *Let $\{X_n\}_{n=1}^{\infty}$ be a sequence of independent real variables with the same distribution. Suppose that $\varphi(t) = \mathbf{E}(\exp tX_1)$ exists for all t in an open interval I containing 0 and fix a real number $s > \mathbf{E}(X)$ such that $t \mapsto e^{-ts}\varphi(t)$ attains its minimum $\alpha(s)$ at a point τ of I. Let $S_n = X_1 + \cdots + X_n$.*
(1) Show that $\log\varphi(t)$ is convex on I and that $\tau > 0$. Conclude, using Chebyshev's inequality (Problem IV-6), that

$$\left[P\left(\frac{S_n}{n} \geq s\right)\right]^{\frac{1}{n}} \leq \alpha(s).$$

(2) Let μ_1 be the distribution of $X_1 - s$ and let $\nu(dx) = \frac{e^{\tau x}}{\alpha(s)}\mu_1(dx)$. Prove that ν is a probability measure, that $\int x\nu(dx) = 0$, and that $\int x^2\nu(dx) < \infty$.
(3) Let $\{Z_n\}_{n=1}^{\infty}$ be a sequence of independent random variables with the same distribution ν. Show that

$$P\left[\frac{S_n}{n} \geq s\right] = (\alpha(s))^n \mathbf{E}[\exp(-\tau(Z_1 + \cdots + Z_n))\mathbf{1}_{\{Z_1+\cdots+Z_n \geq 0\}}].$$

Conclude from Problem IV-8 that

$$\alpha(s) = \lim_{n\to\infty}\left[P\left(\frac{S_n}{n} \geq s\right)\right]^{\frac{1}{n}}.$$

(4) Compute $\alpha(s)$ in the following cases.

a) $\varphi(t) = \exp(\frac{t^2}{2})$ (normal distribution)

b) $\varphi(t) = \cosh(t)$ (Bernouilli distribution)

c) $\varphi(t) = (1-t)^{-\alpha}$, $t < 1$, $\alpha > 0$ (gamma distribution)

d) $\varphi(t) = \exp\lambda(e^t - 1)$, $\lambda > 0$ (Poisson distribution)

e) $\varphi(t) = (\frac{p}{1-qe^t})$, $p+q=1$, $0<p<1$, $t < -\log q$, $\alpha > 0$ (negative binomial distribution)

f) $\varphi(t) = \frac{1}{1-t^2}$, $|t| < 1$ (Laplace's first distribution)

g) $\varphi(t) = \frac{1}{\cos t}$ (logarithm of a Cauchy distribution)

SOLUTION. (1) It is easy to see that $\varphi^{(n)}(t) = \mathbf{E}(X_1^n \exp(tX_1))$ for every nonnegative integer n if $t \in I$, and that $s = \frac{\varphi'(\tau)}{\varphi(\tau)}$.

$(\log\varphi(t))'' = \frac{1}{\varphi^2(t)}[\varphi''(t)\varphi(t) - \varphi^2(t)]$. However, introducing the probability measure $\gamma(dx) = \frac{e^{tx}}{\varphi(t)}\mu(dx)$, we find that

$$(\log\varphi(t))'' = \int\int(x^2 - xy)\gamma(dx)\gamma(dy),$$

which is positive. (This is the classical inequality $\mathbf{E}((X - \mathbf{E}(X))^2) = \mathbf{E}(X^2) - (\mathbf{E}(X))^2 \geq 0$ for a random variable X with distribution γ.) Hence $t \mapsto \frac{\varphi'(t)}{\varphi(t)}$ is increasing on I; since $s > \mathbf{E}(X_1) = \frac{\varphi'(0)}{\varphi(0)}$, this implies that $\tau > 0$.

Chebyshev's inequality then gives

$$P\left[\frac{S_n}{n} \geq s\right] = P[\exp(\tau S_n) \geq \exp(\tau s n)] \leq \left(e^{-\tau_s}\varphi(\tau)\right)^n = (\alpha(s))^n.$$

(2) $\int\nu(dx) = \frac{1}{\alpha(s)}\mathbf{E}[\exp\tau(X_1 - s)] = 1$ by definition. Since τ lies in I,

$$\int|x|^n\nu(dx) = \frac{1}{\alpha(s)}\mathbf{E}(|X_1 - s|^n\exp(\tau(X_1 - s)) < \infty \quad \text{for all } n.$$

Since $s = \frac{\varphi'(\tau)}{\varphi(\tau)}$,

$$\int x\nu(dx) = \frac{e^{-\tau_s}}{\alpha(s)}\mathbf{E}((X_1 - s)\exp(\tau X_1)) = 0.$$

$$\begin{aligned}(3)\ P[\tfrac{S_n}{n} \geq s] &= \int_{y_1+\cdots+y_n\geq 0}\mu_1(dy_1)\ldots\mu_1(dy_n)\\ &= (\alpha(s))^n\int_{y_1+\cdots+y_n\geq 0} e^{-\tau(y_1+\cdots+y_n)}\nu(dy_1)\ldots\nu(dy_n)\\ &= (\alpha(s))^n\mathbf{E}(\exp(-\tau(Z_1+\cdots+Z_n))\mathbf{1}_{\{Z_1+\cdots+Z_n\geq 0\}}).\end{aligned}$$

Applying Problem IV-8 to the independent variables $\tau Z_1, \ldots, \tau Z_n$ now gives the desired result.
(4) The computations are standard and give

$$\begin{array}{llll}
a) & \alpha(s) & = \exp(-\frac{s^2}{2}) & s>0 \\
b) & \alpha(s) & = \frac{1}{2}\left[\left(\frac{1+s}{1-s}\right)^{\frac{1-s}{2s}} + \left(\frac{1+s}{1-s}\right)^{-\frac{1+s}{2s}}\right] & 0<s<1 \\
c) & \alpha(s) & = e^{-s+\alpha}\left(\frac{s}{\alpha}\right)^{\alpha} & s>\alpha \\
d) & \alpha(s) & = e^{s-\lambda}\left(\frac{\lambda}{s}\right)^{s} & s>\lambda \\
e) & \alpha(s) & = p^{\alpha}(1+\frac{s}{\alpha})^{\alpha}\left[\frac{s}{q(\alpha+s)}\right]^{-s} & s>\alpha\cdot\frac{q}{p} \\
f) & \alpha(s) & = \exp(1-\sqrt{1+s^2})\frac{s^2}{2(\sqrt{1+s^2}-1)} & s>0 \\
g) & \alpha(s) & = \exp(-s\ \arctan\ s)(1+s^2)^{\frac{1}{2}} & s>0.
\end{array}$$

REMARK. It is not known what conditions on a decreasing function α on $\mathbf{R}$ are sufficient for the existence of a distribution μ of the X_n such that

$$\lim_{n\to\infty}\left[P\left(\frac{S_n}{n}\geq s\right)\right]^{\frac{1}{n}} = \alpha(s).$$

Problem IV-11. *If z_1 and z_2 are complex numbers with positive real part, set $\Gamma(z_1) = \int_0^\infty x^{z_1-1}e^{-x}dx$ and $B(z_1,z_2) = \int_0^1 x^{z_1-1}(1-x)^{z_2-1}dx$. Assume without proof the formula*

$$B(z_1,z_2) = \frac{\Gamma(z_1)\Gamma(z_2)}{\Gamma(z_1+z_2)}.$$

If a and b are positive, the probability measures

$$\gamma_a(dx) = \mathbf{1}_{(0,+\infty)}(x)x^{a-1}e^{-x}\frac{dx}{\Gamma(a)}$$
$$\beta_{a,b}(dx) = \mathbf{1}_{(0,1)}(x)x^{a-1}(1-x)^{b-1}\frac{dx}{B(a,b)}$$
$$\beta^{(2)}_{a,b}(dx) = \mathbf{1}_{(0,+\infty)}(x)x^{a-1}(1+x)^{-a-b}\frac{dx}{B(a,b)}$$

are called, respectively, the gamma distribution with parameter a and the beta distributions of the first and second kind with parameters a and b.

(1) If μ is a bounded measure on $(0,+\infty)$, its Mellin transform is $(M\mu)(t) = \int_0^\infty x^{it}\mu(dx)$ for t real. (This is the Fourier transform of the image of μ under $x\mapsto \log x$.) Compute $M\gamma_a$, $M\beta_{ab}$, and $M\beta^{(2)}_{ab}$.

(2) If X is a random variable with distribution $\beta_{a,b}$, compute the distribution of $\frac{X}{1-X}$.

(3) If X and Y are independent r.v. with distributions γ_a and γ_b, compute the distributions of $\frac{X}{Y}$ and $\frac{X}{X+Y}$.

(4) If X, Y, and Z are independent r.v. with distributions $\beta_{a,b}$, $\beta_{a+b,c}$, and $\beta^{(2)}_{a+b.c}$, compute the distributions of XY and XZ.

SOLUTION. (1) It is immediate that $(M_{\gamma_a})(t) = \frac{\Gamma(a+it)}{\Gamma(a)}$ and $(M\beta_{a,b})(t) = \frac{B(a+it,b)}{B(a,b)}$. For $M\beta^{(2)}_{a,b}$, the change of variable $u = \frac{1}{1+x}$ in the integral gives

$$(M\beta^{(2)}_{a,b})(t) = \frac{B(a+it, a-it)}{B(a,b)}.$$

(2) $\mathbf{E}[X^{it}(1-X)^{-it}] = (M\beta^{(2)}_{ab})(t)$ by (1). Hence $\frac{X}{1-X}$ has distribution $\beta^{(2)}_{a,b}$.
(3) $\mathbf{E}[X^{it}Y^{-it}] = \mathbf{E}[X^{it}]\mathbf{E}[Y^{-it}]$ by the independence of X and Y.

(1) and the relation between B and Γ imply that $\mathbf{E}[X^{it}Y^{-it}] = (M\beta^{(2)}_{a,b})(t)$. Therefore $\frac{X}{Y}$ has distribution $\beta^{(2)}_{a,b}$: if $Z = \frac{X}{(X+Y)}$, then $\frac{X}{Y} = \frac{Z}{1-Z}$. By (2), it follows that the distribution of Z is $\beta_{a,b}$.
(4) Similarly, $\mathbf{E}[X^{it}Y^{it}] = \mathbf{E}[X^{it}]\mathbf{E}[Y^{it}] = (M\beta_{a,b+c})(t)$ and $\mathbf{E}[X^{it}Z^{it}] = (M\beta^{(2)}_{a,c})(t)$.

Problem IV-12. *(1) Let γ_a be the probability measure of Problem IV-11, with $a > 0$. Compute its Fourier transform. If X and Y are independent random variables with distributions γ_a and γ_b, compute the distribution of $X+Y$.*
(2) Let X be a Gaussian random variable with density $\frac{1}{\sigma\sqrt{2\pi}} e^{-\frac{x^2}{2\sigma^2}}\,dx$. Compute $\mathbf{E}\left[\left(\frac{X^2}{2\sigma^2}\right)^{it}\right]$ for t real, and use Problem IV-11 to find the distribution of $\frac{X^2}{2\sigma^2}$.
(3) Let $X_1, \ldots, X_d$, $Y_1, \ldots, Y_m$ be independent random variables with the same distribution as X of (2). Compute the distribution of $\frac{1}{2\sigma^2}[X_1^2 + \ldots + X_d^2]$ by using (1) and (2), and the distributions of

$$\frac{X_1^2 + \cdots + X_d^2}{Y_1^2 + \cdots + Y_m^2} \quad \text{and} \quad \frac{X_1^2 + \cdots + X_d^2}{X_1^2 + \cdots + X_d^2 + Y_1^2 + \cdots + Y_m^2}$$

by using Problem IV-11(3).

SOLUTION. $\widehat{\gamma}_a(t) = \int_0^\infty e^{x(it-1)} x^{a-1} \frac{dx}{\Gamma(a)}$ is computed by integrating the analytic function $z^{a-1}e^{-z}$ along the contour shown in the figure; for Re $z > 0$, we take the value which is real and positive on $(0, +\infty)$. BC and DA are arcs of circles centered at 0, with radii R and ϵ, and CD passes

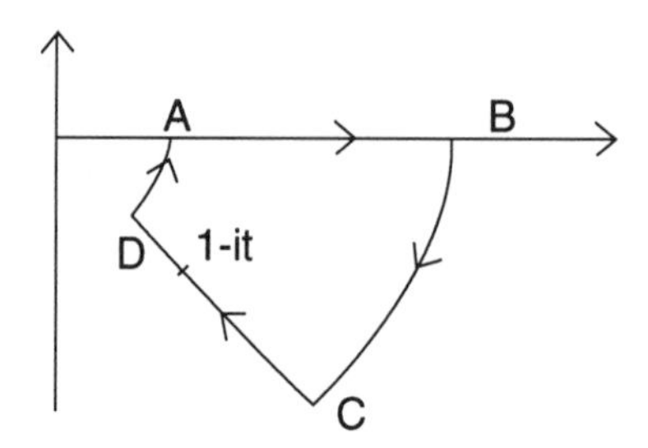

through the point with complex coordinate $1 - it$. By Cauchy's theorem, $0 = \int_A^B + \int_B^C + \int_C^D + \int_D^A$. On CD, the change of variable $z = (1-it)x$ gives

$$\int_C^D = -\int_{\frac{\epsilon}{\sqrt{1+t^2}}}^{\frac{R}{\sqrt{1+t^2}}} e^{x(it-1)}(1-it)^a x^{a-1} dx.$$

(Recall that the branch of z^a chosen makes $(1-it)^a$ unambiguous for $t \in \mathbf{R}$). Clearly $a > 0$ implies $|\int_D^A| \leq \epsilon^a \to 0$ as $\epsilon \to 0$. The complex coordinate of C is $\frac{R}{\sqrt{1+t^2}}(1-it)$; hence $\operatorname{Re} z \geq \frac{R}{\sqrt{1+t^2}}$ as z traverses BC. Hence

$$\left|\int_B^C\right| \leq R^a(\arctan\ t)\exp\left(-\frac{R}{\sqrt{1+t^2}}\right) \to 0 \quad \text{as } R \to +\infty.$$

Thus $\lim_{\substack{R\to+\infty \\ \epsilon\to 0}} \int_A^B + \int_C^D = 0$ and $\widehat{\gamma}_a(t) = (1-it)^{-a}$. It follows that

$$\mathbf{E}[\exp(it(X+Y))] = \widehat{\gamma}_a(t)\widehat{\gamma}_b(t) = \widehat{\gamma}_{a+b}(t)$$

and that $X + Y$ has distribution γ_{a+b}.

(2)

$$\mathbf{E}\left[\left(\frac{X^2}{2\sigma^2}\right)^{it}\right] = \frac{1}{\sigma\sqrt{2\pi}}\int_{-\infty}^{+\infty}\left(\frac{x^2}{2\sigma^2}\right)^{it} e^{-\frac{x^2}{2\sigma^2}}dx = \int_0^\infty u^{it}u^{-\frac{1}{2}}e^{-u}\frac{du}{\sqrt{\pi}},$$

where the last equality follows from the change of variable $u = \frac{x^2}{2\sigma^2}$. Since $B(1,1) = \Gamma(1) = 1$ trivially, the relation between B and Γ given in Problem IV-11 shows that $\Gamma(\frac{1}{2}) = \sqrt{\pi}$; thus $\frac{X^2}{2\sigma^2}$ has distribution $\gamma_{\frac{1}{2}}$ by (1) of Problem IV-11.

(3) It is immediate that $\gamma_{\frac{d}{2}}$ is the distribution of $\frac{1}{2\sigma^2}[X_1^2 + \cdots + X_d^2]$ and that $\beta_{\frac{d}{2},\frac{m}{2}}$ is the distribution of the other two.

Problem IV-13. *In Euclidean space $\mathbf{R}^{d+1}$, consider a random variable $X = (X_0, X_1, \ldots, X_d)$ whose distribution μ is invariant under every orthogonal matrix of $\mathbf{R}^{d+1}$ and satisfies $\mu(\{0\}) = 0$. Let ν denote the distribution of X on $(0, +\infty)$ and let $Y = (\frac{X_1}{X_0}, \frac{X_2}{X_0}, \ldots, \frac{X_d}{X_0})$.*

(1) Use Problem III-2 to show that the distribution of Y is independent of ν.
(2) From now on, assume that the $\{X_j\}_{j=0}^n$ are independent, with the same distribution μ and with Fourier transform $\exp(-\frac{t^2}{2})$. Show, using Problem III-1, that μ must be invariant under every orthogonal matrix.
(3) If $a \in \mathbf{R}$, compute the integral

$$I(a) = \frac{1}{\sqrt{2\pi}} \int_{-\infty}^{+\infty} \exp\left[-\frac{1}{2}\left(x^2 + \frac{a^2}{x^2}\right)\right] dx$$

by using the following fact from Problem II-12(1):

$$\int_{-\infty}^{+\infty} f\left(x - \frac{|a|}{x}\right) dx = \int_{-\infty}^{+\infty} f(y)dy \quad \text{for every } f \text{ integrable on } \mathbf{R}.$$

(4) By first conditioning with respect to X_0 (see Problem IV-34), compute the Fourier transform of the distribution of Y.
(5) Using Problem IV-11, find the distribution of $\|Y\|^2$. Derive the density of Y from this, by observing that the distribution of Y is invariant under every orthogonal matrix in O_d and using Problem III-3.

SOLUTION. (1) By Problem III-2(4), if $U = (U_0, \dots, U_d)$ is a random variable concentrated on the unit sphere S_d of $\mathbf{R}^{d+1}$, independent of $\|X\|$ and with rotation-invariant distribution σ, then X and $U\|X\|$ have the same distribution. Hence Y and $(\frac{U_1}{U_0}, \frac{U_2}{U_0}, \dots, \frac{U_d}{U_0})$ have the same distribution, which proves the result.
(2) $\mathbf{E}[\exp(i\sum_{j=0}^d t_j X_j)] = \exp(-\frac{1}{2}\|t\|^2)$. Problem III-1 gives the result.
(3) Set $f(y) = \frac{1}{\sqrt{2\pi}}\exp(-\frac{y^2}{2})$. Then

$$f\left(x - \frac{|a|}{x}\right) = e^{+|a|} \exp\left(-\frac{1}{2}\left(x^2 + \frac{a^2}{x^2}\right)\right).$$

Hence $I(a) = e^{-|a|} I(0) = e^{-|a|}$.

$$(4) \qquad \mathbf{E}\left[\exp\left(i\sum_{j=1}^d t_j \frac{X_j}{X_0}\right)\right] = \mathbf{E}\left[\mathbf{E}\left[\exp\left(i\sum_{j=1}^d t_j \frac{X_j}{X_0}\right) \mid X_0\right]\right].$$

But, by (2),

$$\mathbf{E}\left[\exp\left(i\sum_{j=1}^d t_j \frac{X_j}{X_0}\right) \mid X_0\right] = \exp\left(-\frac{\|t\|^2}{2X_0^2}\right).$$

Using (3), the Fourier transform of the distribution of Y is thus $\exp(-\|t\|)$.

(5) By Problem IV-11, $\|Y\|^2$ has distribution $\beta^{(2)}_{\frac{d}{2},\frac{1}{2}}$; that is, the density on $(0,+\infty)$ equals $\dfrac{x^{\frac{d}{2}-1}}{B(\frac{d}{2},\frac{1}{2})(1+x)^{\frac{d+1}{2}}}$. The density f of Y on $\mathbf{R}^d$ exists and is continuous (for example, because its Fourier transform $\exp(-\|t\|)$ is integrable by the Fourier inversion theorem). Moreover, since the distribution of Y is rotation invariant, its density is of the form $f_1(\|x\|)$, where f_1 is the density of the distribution of $\|Y\|$ with respect to the measure ν_0 on $(0,+\infty)$ (see Problem III-3). We thus differentiate the following equation with respect to r:

$$\begin{aligned} P[\|Y\|<r]=P[\|Y\|^2<r^2] &= \int_0^{r^2} \frac{x^{\frac{d}{2}-1}}{(1+x)^{\frac{d+1}{2}}}\frac{dx}{B(\frac{d}{2},\frac{1}{2})} \\ &= \frac{2\pi^{\frac{d}{2}}}{\Gamma(\frac{d}{2})}\int_0^{r} f_1(\rho)\rho^{d-1}d\rho. \end{aligned}$$

Using the relation between B and Γ (Problem IV-11) and the fact that $\Gamma(\frac{1}{2})=\sqrt{\pi}$, we finally obtain the density of Y:

$$f(x)=\frac{\Gamma(\frac{d+1}{2})}{\pi^{\frac{d+1}{2}}}\frac{1}{(1+\|x\|^2)^{\frac{d+1}{2}}}.$$

Problem IV-14. *Let γ_a be the probability measure of Problem IV-11, with $a>0$.*
(1) Use Problem IV-12 to compute $\lim_{a\to\infty}\int_0^\infty \exp[it(\frac{x-a}{\sqrt{a}})]\gamma_a(dx)$.
(2) Using Problem II-14, show that

$$\lim_{a\to+\infty}\int_0^\infty \left|\frac{x-a}{\sqrt{a}}\right|\gamma_a(dx)=\int_{-\infty}^{+\infty}|x|e^{-\frac{x^2}{2}}\frac{dx}{\sqrt{2\pi}}.$$

(3) Integrate by parts to compute the integral $\int_0^\infty |\frac{x-a}{\sqrt{a}}|\gamma_a(dx)$ and prove Stirling's formula:

$$\lim_{a\to+\infty}\frac{a^{a-\frac{1}{2}}e^{-a}}{\Gamma(a)}=\frac{1}{\sqrt{2\pi}}.$$

SOLUTION. (1) With the convention, as in Problem IV-12, of choosing the branch of z^a on $\{z : \text{Re } z>0\}$ which is real and positive on $(0,+\infty)$, we have

$$\int_0^\infty \exp\left(it(\frac{x-a}{\sqrt{a}})\right)\gamma_a(dx)=(1-\frac{it}{\sqrt{a}})^{-a}\exp(-it\sqrt{a}).$$

Setting $h=\frac{1}{\sqrt{a}}$ and expanding the logarithm of the complex function on the right-hand side in a power series with respect to h (with the convention above for determining z^a), we obtain

$$\log \int_0^{\infty} \exp\left(it(\frac{x-a}{\sqrt{a}})\right) \gamma_a(dx) = -\frac{it}{h} + \frac{1}{h^2}\sum_{k=1}^{\infty} \frac{(it)^k}{k} h^k = -\frac{t^2}{2} + h\sum_{k=3}^{\infty} \frac{(it)^k}{k} h^{k-3}.$$

As $h \to 0$, the right-hand side approaches $-\frac{t^2}{2}$. But this implies that if X_a is a random variable with density γ_a, then the limit distribution of μ_a, the distribution of $\frac{(X_a - a)}{\sqrt{a}}$, has density $\frac{1}{\sqrt{2\pi}} \exp(-\frac{x^2}{2})dx$.
(2) Problem II-14(2) is applicable since

$$\int_0^{\infty} (\frac{x-a}{\sqrt{a}})^2 \gamma_a(dx) = \int_{-\infty}^{+\infty} x^2 \mu_a(dx) = 1 \quad \text{for all } a > 0.$$

(3) It follows easily from $\int x^a \mathrm{e}^{-x} dx = -x^a \mathrm{e}^{-x} + a \int x^{a-1} \mathrm{e}^{-x} dx$ that

$$\int_0^{\infty} |x-a| x^{a-1} \mathrm{e}^{-x} dx = 2a^a \mathrm{e}^{-a},$$

by splitting the integral into $\int_0^a$ and $\int_a^{+\infty}$. (2) then implies that

$$\int_0^{\infty} |\frac{x-a}{\sqrt{a}}| \gamma_a(dx) = \frac{2a^{a-\frac{1}{2}} \mathrm{e}^{-a}}{\Gamma(a)}.$$

But, as $a \to \infty$, this approaches

$$\int_{-\infty}^{+\infty} |x| \mathrm{e}^{-\frac{x^2}{2}} \frac{dx}{\sqrt{2\pi}} = \frac{2}{\sqrt{2\pi}}.$$

REMARK. The same type of calculation is feasible for the Poisson distribution, since $\sum_{k=0}^{\infty} |k - \lambda| \frac{\lambda^k}{k!} = 2\frac{\lambda^{[\lambda]}}{[\lambda]!}$, where $[\lambda]$ is the greatest integer $\leq \lambda$ when $\lambda > 0$.

Problem IV-15. *(1) Let μ be a probability measure on $\mathbf{R}$ such that $\widehat{\mu}(t) = \widehat{\mu}(t\cos\theta)\widehat{\mu}(t\sin\theta)$ for all real t and θ. Show that there exists $\sigma \geq 0$ such that $\widehat{\mu}(t) = \exp(-\frac{\sigma^2 t^2}{2})$.*
METHOD. Show that $\widehat{\mu}(t) \geq 0$, then that $\widehat{\mu}(t) > 0$ for every t. Finally, consider $f(x) = -\log \widehat{\mu}(\sqrt{x})$ for $x \geq 0$.

(2) For positive integers d_1 and d_2, let μ_1 and μ_2 be probability measures on the Euclidean spaces $\mathbf{R}^{d_1}$ and $\mathbf{R}^{d_2}$ such that $\nu = \mu_1 \otimes \mu_2$ is invariant under the group G of orthogonal matrices on $\mathbf{R}^{d_1+d_2}$. Show that there exists

$\sigma \geq 0$ *such that* $\widehat{\mu}_j(t) = \exp(-\frac{\sigma^2}{2}\|t\|_j)$, $j = 1, 2$, *where* $\|t\|_1$ *and* $\|t\|_2$ *are the norms in* $\mathbf{R}^{d_1}$ *and* $\mathbf{R}^{d_2}$.

METHOD. Use Problem III-1 and part (1) of this problem for the case where $d_1 = d_2 = 1$.

SOLUTION. (1) $\theta = -\frac{3\pi}{4}$ gives $\widehat{\mu}(t) = \widehat{\mu}(-\frac{t}{\sqrt{2}})\widehat{\mu}(\frac{t}{\sqrt{2}}) = |\widehat{\mu}(-\frac{t}{\sqrt{2}})|^2 \geq 0$. If there were a real t_0 such that $\widehat{\mu}(t_0) = 0$, the preceding inequality would imply $\widehat{\mu}(2^{-\frac{n}{2}}t_0) = 0$ for every integer $n > 0$. Since $\widehat{\mu}$ is continuous, we would obtain the following contradiction: $1 = \widehat{\mu}(0) = \lim_{n\to\infty} \widehat{\mu}(2^{-\frac{n}{2}}t_0) = 0$. Therefore the range of $\widehat{\mu}$ is contained in $(0,1]$ and $f(x) = -\log \mu(\sqrt{x}) \geq 0$. Since $f(x) = f(x\cos^2\theta) + f(x\sin^2\theta)$ for all $x \geq 0$ and all θ in $\mathbf{R}$, f satisfies the functional equation

$$f(x+y) = f(x) + f(y) \quad \text{for all } x,\, y \geq 0.$$

It follows easily that, if $f(1) = \frac{\sigma^2}{2} \geq 0$ and x is nonnegative and rational, $f(x) = \frac{x\sigma^2}{2}$. Since f is continuous on $[0, +\infty)$ this is true even for irrational positive x, which completes the proof.
(2) Let $d_1 = d_2 = 1$ and let

$$\widehat{\nu}(t_1, t_2) = \int_{\mathbf{R}^2} \exp\ i(x_1t_1 + x_2t_2)\mu_1(dx_1)\mu_2(dx_2) = \widehat{\mu}_1(t_1)\widehat{\mu}_2(t_2).$$

Then, by Problem III-1, for all $t \in \mathbf{R}$ the map $\theta \mapsto \widehat{\nu}(t\cos\theta, t\sin\theta)$ is independent of θ. Setting θ equal first to 0 and then to $\frac{\pi}{2}$, we find that for all $t \in \mathbf{R}$

$$\widehat{\mu}_1(t\cos\theta)\widehat{\mu}_2(t\sin\theta) = \widehat{\mu}_1(t) = \widehat{\mu}_2(t),$$

and the result follows from (1).

For arbitrary d_1 and d_2, take a_1 and a_2 on the unit spheres of $\mathbf{R}^{d_1}$ and $\mathbf{R}^{d_2}$, respectively. Then $(a_1\cos\theta, a_2\sin\theta)$ lies on the unit sphere of $\mathbf{R}^{d_1+d_2}$ for all real θ. Let ν_1 and ν_2 denote the images in $\mathbf{R}$ of μ_1 and μ_2 under the mappings $x_1 \mapsto \langle a_1, x_1\rangle$ and $x_2 \mapsto \langle a_2, x_2\rangle$. Then Problem III-1 implies that ν_1 and ν_2 are independent of a_1 and a_2, and the case $d_1 = d_2 = 1$ already considered implies the existence of $\sigma \geq 0$ such that $\widehat{\nu_1}(t) = \widehat{\nu_2}(t) = \exp(-\frac{\sigma^2}{2}t^2)$. Problem III-1 then implies that $\widehat{\mu}_j = \exp(-\frac{\sigma^2}{2}\|t\|_j)$ for $t \in \mathbf{R}^{d_j}$.

REMARK. The converse of the property in (2) is trivial. This characterization of centered normal distributions is sometimes called Maxwell's theorem.

Problem IV-16. *A real random variable* Z *is called* symmetric *if* Z *and* $-Z$ *have the same distribution.*

(1) Show that Z has distribution $c(dz) = \frac{dz}{\pi(1+z^2)}$ if and only if Z is symmetric and $|Z|^2$ has distribution $\beta^{(2)}(\frac{1}{2}, \frac{1}{2})$. (See Problem IV-11.) Assuming without proof the formula $\Gamma(z)\Gamma(1-z) = \frac{\pi}{\sin \pi z}$ for complex numbers z such that $0 < Re z < 1$, compute $\mathbf{E}(|Z|^{it})$ for real t in this case.
(2) Let X_1 and X_2 be two real random variables which are independent and symmetric, and have distributions μ_1 and μ_2 such that $\mu_1(\{0\}) = \mu_2(\{0\}) = 0$. Show that $Z = \frac{X_2}{X_1}$ has distribution c in the following cases:

(a) $\mu_1(dx) = \mu_2(dx) = \exp(-\frac{x^2}{2})\frac{dx}{\sqrt{2\pi}}$

(b) $|X_1|^2$ has distribution $\beta(\frac{1}{2}, b)$ and $|X_2|^2$ has distribution $\beta^{(2)}(\frac{1}{2}, \frac{1}{2}+b)$

(c) $\mu_1(dx) = \mu_2(dx) = \frac{\sqrt{2}}{\pi}\frac{dx}{1+x^4}$

(3) With X_1 and X_2 as in (2), deduce from (2a) that $U = (\frac{X_1}{\sqrt{X_1^2+X_2^2}}, \frac{X_2}{\sqrt{X_1^2+X_2^2}})$ is uniformly distributed on the unit circle of Euclidean space $\mathbf{R}^2$ if and only if $Z = \frac{X_2}{X_1}$ has distribution c.

SOLUTION. (1) If Z has distribution c, Z is clearly symmetric. Moreover,

$$\mathbf{E}[|Z|^{2it}] = \frac{2}{\pi}\int_0^\infty \frac{|z|^{2it}}{1+z^2}dz = \frac{1}{B(\frac{1}{2}, \frac{1}{2})}\int_0^\infty \frac{u^{-\frac{1}{2}+it}}{1+u}du,$$

by the change of variable $u = z^2$. The converse is immediate.

By the preceding computation, Problem IV-11, and the formula given above for the gamma function,

$$\mathbf{E}[|Z|^{it}] = \frac{1}{\pi}\Gamma(\frac{1}{2}+\frac{it}{2})\Gamma(\frac{1}{2}-\frac{it}{2}) = \frac{1}{\sin\pi(\frac{1}{2}+\frac{it}{2})} = \frac{1}{\cosh\frac{\pi t}{2}}\frac{2}{\pi}\int_0^\infty \frac{u^{it}\,du}{1+u^2}.$$

(2)

(a)

$$\begin{aligned}\mathbf{E}\left[\left|\frac{X_2}{X_1}\right|^{it}\right] &= \frac{4}{2\pi}\int_0^\infty\int_0^\infty \left|\frac{x_2}{x_1}\right|^{it}\exp\left(-\frac{1}{2}(x_1^2+x_2^2)\right)dx_1dx_2\\ &= \frac{2}{\pi}\int_0^\infty \rho e^{-\frac{\rho^2}{2}}d\rho\int_0^{\frac{\pi}{2}}(\tan\theta)^{it}d\theta,\end{aligned}$$

where we have switched to polar coordinates. Setting $u = \tan\theta$, we obtain

$$\mathbf{E}\left[\left|\frac{X_2}{X_1}\right|^{it}\right] = \frac{2}{\pi}\int_0^\infty \frac{u^{it}\,du}{1+u^2} = \frac{1}{\cosh\frac{\pi t}{2}}.$$

Since $\frac{X_2}{X_1}$ is symmetric, (1) shows that it has distribution c.

(b)

$$\begin{aligned}\mathbf{E}\left[|X_2|^{2it}|X_1|^{-2it}\right] &= \frac{\Gamma(\frac{1}{2}+it)\Gamma(\frac{1}{2}+b-it)}{\Gamma(\frac{1}{2})\Gamma(\frac{1}{2}+b)} \times \frac{\Gamma(\frac{1}{2}-it)\Gamma(\frac{1}{2}+b)}{\Gamma(\frac{1}{2}+b-it)\Gamma(\frac{1}{2})}\\ &= \frac{1}{\pi}\Gamma(\frac{1}{2}+it)\Gamma(\frac{1}{2}-it)\end{aligned}$$

Hence $|\frac{X_2}{X_1}|^2$ has distribution $\beta^{(2)}(\frac{1}{2},\frac{1}{2})$, and (1) gives the desired result.
(c) *First method:*

$$\begin{aligned}
\mathbf{E}[|X_2|^{it}|X_1|^{-it}] &= \left|\frac{2\sqrt{2}}{\pi}\int_0^\infty \frac{x^{it}}{1+x^4}dx\right|^2 \\
&= \left|\frac{1}{B(\frac{1}{4},\frac{3}{4})}\int_0^\infty \frac{u^{-\frac{3}{4}+\frac{it}{4}}}{1+u}du\right|^2 = \left|\frac{\Gamma(\frac{1}{4}+\frac{it}{4})\Gamma(\frac{3}{4}-\frac{it}{4})}{\Gamma(\frac{1}{4})\Gamma(\frac{3}{4})}\right|^2 \\
&= \left|\frac{1}{\sqrt{2}\sin\pi(\frac{1}{4}+it)}\right|^2 = \frac{1}{\cosh\frac{\pi t}{2}}
\end{aligned}$$

Second method:

$$\begin{aligned}
P[X_2 < tX_1] = F(t) &= \int\int_{y<tx} \frac{2}{\pi^2}\frac{dx\,dy}{(1+x^4)(1+y^4)} \\
&= \frac{2}{\pi^2}\int_{v<t} dv \int_{-\infty}^{+\infty} \frac{|u|du}{(1+u^4)(1+(uv)^4)},
\end{aligned}$$

with the change of variables $x = u$ and $y = uv$. Setting $w = u^2$ gives

$$F'(t) = \frac{2}{\pi^2}\int_0^\infty \frac{dw}{(1+w^2)(1+w^2v^4)} = \frac{1}{\pi(1+v^2)},$$

using the decomposition

$$\frac{1}{(1+w^2)(1+w^2v^4)} = \frac{1}{1-v^4}\left[\frac{1}{1+w^2} - \frac{1}{\frac{1}{v^4}+w^2}\right].$$

(3) The fact that X_1 and X_2 are symmetric implies that the distribution of U does not depend on the distribution of $\frac{X_1}{X_2}$. If X_1 and X_2 are as in (2a) then, since the distribution of (X_1, X_2) is rotation invariant, U is uniform. But in this case $\frac{X_1}{X_2}$ has distribution c; we conclude that, in general, U is uniform whenever $\frac{X_1}{X_2}$ has distribution c. The converse is immediate: if U is uniform, then $\frac{X_1}{X_2}$ necessarily satisfies

$$\mathbf{E}\left[|\frac{X_1}{X_2}|^{it}\right] = \frac{4}{2\pi}\int_0^{\frac{\pi}{2}}(\tan\theta)^{it}d\theta = \frac{1}{\cosh\frac{\pi t}{2}}.$$

That is, $\frac{X_1}{X_2}$ has distribution c.

REMARKS. Example (2c) is due to Laha (1949). Moreover, if (X_1, X_2) is as in (2) with U uniform, then $(\frac{1}{X_1},\frac{1}{X_2})$ has the same property.

Problem IV-17. *A probability measure ν on a Euclidean space $\mathbf{R}^d$ is called* isotropic *if $\nu(\{0\}) = 0$ and the image of ν under the mapping $x \mapsto \frac{x}{\|x\|}$, in the unit sphere S_{d-1} of $\mathbf{R}^d$, is the unique rotation-invariant probability* *measure σ_{d-1} on S_{d-1}. It is called* radial *if its image ν_a in $\mathbf{R}$ under the mapping $x \mapsto \langle a, x\rangle$ does not depend on a when a ranges over the unit sphere.*

(1) Let μ_1 and μ_2 be probability measures on the Euclidean spaces $\mathbf{R}^{d_1}$ and $\mathbf{R}^{d_2}$, with d_1 and d_2 positive. Show that the probability measure $\nu = \mu_1 \otimes \mu_2$ on the Euclidean space $\mathbf{R}^{d_1+d_2}$ is isotropic if and only if μ_1 and μ_2 are radial and if, for every a_1 in S_{d_1-1} and a_2 in S_{d_2-1}, the image of ν under $(x_1, x_2) \mapsto \frac{\langle a_2, x_2\rangle}{\langle a_1, x_1\rangle}$ is $c(dz) = \frac{dz}{\pi(1+z^2)}$.

METHOD. Prove the assertion first for $d_1 = d_2 = 1$ and use Problem IV-16.

(2) Let (X_1, X_2, X_3) be three independent random variables such that the distribution ν of (X_1, X_2, X_3) in $\mathbf{R}^3$ is isotropic. Show that there exists $\sigma > 0$ such that

$$\mathbf{E}[\exp(itX_j)] = \exp\left(-\frac{\sigma^2 t^2}{2}\right) \quad \text{for } j = 1,\ 2,\ 3, \text{ and } t \in \mathbf{R}.$$

METHOD. Apply (1) to the distributions μ_1 of X_1 and μ_2 of (X_2, X_3) and use Problem IV-14.

SOLUTION. (1) Suppose that $d_1 = d_2 = 1$ and that ν is isotropic. It is clear that the image of ν under $(x_1, x_2) \mapsto \frac{x_2}{x_1}$ is c. This implies that $\mu_1(\{0\}) = \mu_2(\{0\}) = 0$. We first show that μ_1 and μ_2 are symmetric. Let μ_1^+ and μ_2^+ be the restrictions of the measures μ_1 and μ_2 to $(0, +\infty)$, and let μ_1^- and μ_2^- be the restrictions of the images of μ_1 and μ_2 under $x \mapsto -x$ to $(0, +\infty)$. Since ν is isotropic we can write, for $\epsilon_1 = \pm 1$, $\epsilon_2 = \pm 1$, and t real,

$$\frac{1}{2\pi}\int_0^{\frac{\pi}{2}} (\tan\theta)^{it} d\theta = \frac{1}{4\cosh\frac{\pi t}{2}} = \int_0^\infty \int_0^\infty x_2^{it} x_1^{-it} \mu_1^{\epsilon_1}(dx_1)\mu_2^{\epsilon_2}(dx_2).$$

Since $\frac{1}{4\cosh\frac{\pi t}{2}}$ is never zero, it follows that $\int_0^\infty x^{it}\mu_j^\epsilon(dx)$ is never zero for $\epsilon = \pm 1$ and $j = 1, 2$; hence

$$\int_0^\infty x^{it}\mu_j^+(dx) = \int_0^\infty x_j^{it}\mu_j^-(dx) \quad \text{for all } t \text{ and for } j = 1, 2.$$

By the uniqueness of the Fourier transform, this shows that $\mu_1^+ = \mu_1^-$ and $\mu_2^+ = \mu_2^-$, which is the desired symmetry.

Next, suppose that d_1 and d_2 are arbitrary positive numbers and that ν is isotropic. Then the image measure $\alpha_{a_1} \otimes \beta_{a_2}$ of ν in $\mathbf{R}^2$ under the mapping $(x_1, x_2) \mapsto (\langle a_1, x_1\rangle, \langle a_2, x_2\rangle)$, where $(a_1, a_2) \in S_{d_1-1} \times S_{d_2-1}$, is itself isotropic. α_{a_1} and β_{a_2} are symmetric by the case $d_1 = d_2 = 1$. Moreover,

$$\int_{-\infty}^{+\infty} |x_2|^{it}\beta_{a_2}(dx_2) \int_{-\infty}^{+\infty} |x_1|^{-it}\alpha_{a_1}(dx_1) = \frac{1}{\cosh\frac{\pi t}{2}},$$

which shows not only that α_{a_1} and β_{a_2} are independent of a_1 and a_2, respectively, but also that μ_1 and μ_2 are radial. Finally, it is clear that the image measure of $\alpha_{a_1} \otimes \beta_{a_2}$ under $(x_1, x_2) \mapsto \frac{x_2}{x_1}$ is c.
(2) μ_2 is radial by (1). Hence, by Problem IV-15, there exists $\sigma \geq 0$ such that $\mathbf{E}[\exp(itX_j)] = \exp(-\frac{\sigma^2 t^2}{2})$ for t real and $j = 2, 3$. Repeating the reasoning above with the distributions μ_1' of (X_1, X_2) and μ_2' of X_3 shows that X_1, X_2, and X_3 have the same distribution. Furthermore, $\sigma \neq 0$ since γ is isotropic, and hence $\nu(\{0\}) = 0$.

REMARKS. The converse of (1) is true but rather lengthy to prove. (2) is true for n independent random variables, $n \geq 3$; this follows easily from the problem. (Problem IV-16 showed that this would be false for $n = 2$.) This property of the normal distribution is due to I. Kotlarski (1966), who proves it with the additional hypothesis that the X_j are symmetric.

Problem IV-18. *Let E be a finite-dimensional real vector space, let E^* be its dual, and let $\langle x, t\rangle$ be the canonical bilinear form on $E \times E^*$. If μ is a probability measure on E, its Fourier transform is defined on E^* by*

$$\widehat{\mu}(t) = \int_E \exp(i\langle x, t\rangle)\mu(dx).$$

(1) If there exists $t_0 \neq 0$ such that $|\widehat{\mu}(t_0)| = 1$, show that μ is concentrated on a countable union of affine hyperplanes and determine them.
METHOD. First consider the case where $\widehat{\mu}(t_0) = 1$.

(2) If there exists a probability measure ν on E such that $\widehat{\mu}(t)\widehat{\nu}(t) = 1$ for every t in E^*, show that μ and ν are Dirac measures.

METHOD. First prove this when $\dim E = 1$.

SOLUTION. (1) If $\widehat{\mu}(t_0) = 1$, then

$$\begin{aligned} 0 &= \int_E (1 - \exp i\langle x, t_0\rangle)\mu(dx) = \int_E \mathrm{Re}(1 - \exp i\langle x, t_0\rangle)\mu(dx) \\ &= \int_E (1 - \cos\langle x, t\rangle)\mu(dx). \end{aligned}$$

Since $1 - \cos\langle x, t_0\rangle \geq 0$, this implies that $\cos\langle x_1, t_0\rangle = 1$ μ-almost everywhere. Hence μ is concentrated on the union $\cup_{k\in\mathbf{Z}} H_k$ of the affine hyperplanes

$$H_k = \{x : \langle x, t_0\rangle = 2k\pi\}, \quad k \in \mathbf{Z}.$$

If $|\mu(t_0)| = 1$ then, since the image of the mapping $E \to \mathbf{C}$, $x \mapsto \exp i\langle x, t_0\rangle$ is the unit circle for nonzero t_0, there exists $x_0 \in E$ such that

$\exp i\langle x_0, t_0\rangle = \widehat{\mu}(t_0)$. Therefore

$$0 = \int_E (1 - \exp i\langle x - x_0, t_0\rangle)\mu(dx),$$

and similarly μ is concentrated on the union of the affine hyperplanes

$$H_k = \{x : \langle x, t_0\rangle = \langle x_0, t_0\rangle + 2k\pi\}, \quad k \in \mathbf{Z}.$$

(2) Since $|\widehat{\mu}(t)| \leq 1$ for all $t \in E^*$, we have $|\widehat{\mu}(t)| = |\widehat{\nu}(t)| = 1$ for all $t \in E^*$. Now, if $E = \mathbf{R}$, the "affine hyperplanes" reduce to points. Hence there exists at least one point a such that $\mu(\{a\}) > 0$, and therefore $\widehat{\mu}(t) = \exp(iat)$, which shows that μ is δ_a. In the general case, let E be given a basis $\{e_1, e_2, \ldots, e_n\}$, let $x \mapsto x_j$ denote projection to the jth component of x, and let μ_j denote the image of μ in $\mathbf{R}$ under the mapping $x \mapsto x_j$. Then, if $(t_1, \ldots, t_n)$ are the components of t in E^* in the dual basis $(e_1^*, \ldots, e_n^*)$, we have $\widehat{\mu}_j(t_j) = \widehat{\mu}(t_j e_j^*)$. Hence $|\widehat{\mu}_j(t_j)| = 1$ for all real t_j, and there exists $a_j \in \mathbf{R}$ such that $\mu_j = \delta_{a_j}$. Since μ is a positive measure, μ is concentrated on the intersection

$$\bigcap_{j=1}^{n} \{x : x_j = a_j\} = \left\{\sum a_j e_j\right\} = \{a\},$$

and $\mu = \delta_a$ and $\nu = \delta_{-a}$.

REMARKS. This result can be generalized by replacing E and E^* by a locally compact abelian group and its group $\widehat{G}$ of continuous characters χ. (See III-1.4.) In this case,

$$\widehat{\mu}(\chi) = \int_G \chi(x)\mu(dx), \quad \chi \in \widehat{G}.$$

If $\widehat{\mu}(\chi_0) = 1$, the same proof shows that μ is concentrated on a closed subgroup of $\widehat{G}$. In general, χ_0 is not surjective on the unit circle T of $\mathbf{C}$. Thus, if $|\widehat{\mu}(\chi_0)| = 1$, the fact that the unit disk $\overline{D}$ is convex must be used in the following way: if ν is the image of μ on T under the mapping $x \mapsto \chi_0(x)$, then $\widehat{\mu}(\chi_0) = \int_T e^{i\theta}\nu(d\theta)$ can be an extremal point of $\overline{D}$ only if ν is concentrated at an extremal point. This ensures the existence of an x_0 in G such that $\chi_0(x_0) = \widehat{\mu}(\chi_0)$.

Problem IV-19. *Let X_1, X_2, Y_1, Y_2 be independent real random variables such that Y_1 and Y_2 are strictly positive and $\mathbf{E}[\exp(itX_j)] = \exp(-\frac{t^2}{2})$ for $j = 1, 2$ and t real. Let $R = [X_1^2Y_1^2 + X_2^2Y_2^2]^{\frac{1}{2}}$. Using Problems IV-16 and IV-18, find the distributions of Y_1 and Y_2 such that $U = (\frac{X_1Y_1}{R}, \frac{X_2Y_2}{R})$ is uniformly distributed on the unit circle of $\mathbf{R}^2$.*

SOLUTION. By Problem IV-16, since X_1Y_1 and X_2Y_2 are symmetric the condition is equivalent to

$$\mathbf{E}\left[\left|\frac{X_2Y_2}{X_1Y_1}\right|^{it}\right] = \frac{1}{\cosh \pi t} \quad \text{for } t \text{ real.}$$

But $\mathbf{E}[|\frac{X_2}{X_1}|^{it}] = \frac{1}{\cosh \pi t}$. Hence $\mathbf{E}[Y_1^{it}Y_1^{-it}] = 1$, or

$$\mathbf{E}[\exp(it \log Y_2)]\mathbf{E}[\exp(-it \log Y_1)] = 1.$$

By Problem IV-18 there exists a real constant a such that $\log Y_2 = a$ and $-\log Y_1 = -a$ with probability 1. Thus Y_1 and Y_2 are equal to the same constant.

Problem IV-20. *Let σ_{d-1} be the uniform probability measure on the unit sphere S_{d-1} of the Euclidean space $\mathbf{R}^d$ and let ν_d be the image of μ_d under the dilation $x \mapsto \sqrt{d}x$.*

Prove that ν_d converges narrowly to $\nu(dx) = \exp(-\frac{x^2}{2})\frac{dx}{\sqrt{2\pi}}$.

METHOD. If $Y_1, \ldots, Y_d, \ldots$ is a sequence of independent random variables with the same distribution ν and if $R_d = [Y_1^2 + \cdots + Y_d^2]^{\frac{1}{2}}$, use the fact that σ_{d-1} is the distribution of $R_d^{-1}(Y_1, Y_2, \ldots, Y_d)$, the weak law of large numbers of Problem IV-6, and Problem I-10.

SOLUTION. ν_d is the distribution of $\frac{\sqrt{d}}{R_d}Y_1$. But $\frac{R_d^2}{d} = \frac{1}{d}[Y_1^2 + \cdots + Y_d^2]$. Therefore, since $\mathbf{E}(Y_1^2) = 1$, for all $\epsilon > 0$ $P[|\frac{R_d^2}{d} - 1| \geq \epsilon] \to 0$ as $d \to \infty$ by Problem IV-6. Applying Problem I-10 to $\Omega = (0, \infty) \times \mathbf{R}$ gives $X_n = (\frac{R_n^2}{n}, Y_1)$ and $f(x, y) = \frac{y}{\sqrt{x}}$. Thus

$$P\left[\left|\frac{\sqrt{d}}{R_d}Y_1 - Y_1\right| \geq \epsilon\right] \to 0 \quad \text{as } n \to \infty,$$

and by IV-1.8.4 this implies that the distribution ν_d of $\frac{\sqrt{d}Y_1}{R_d}$ converges to the distribution ν of Y_1.

REMARK. This property of uniform distributions on spheres is known as Poincaré's lemma.

Problem IV-21. *Let S_n denote the set of probability measures μ on $\mathbf{R}$ such that there exists a probability measure μ_n on the Euclidean space $\mathbf{R}^n$ whose image in $\mathbf{R}$ under $x \mapsto \langle a, x \rangle$ is μ for* every *a in the unit sphere of $\mathbf{R}^n$. Prove that $\mu \in \cap_{n=1}^{\infty} S_n$ if and only if there exists a probabil-*

ity measure ρ on $[0,+\infty)$ such that the Fourier transform of μ satisfies $\widehat{\mu}(t) = \int_0^\infty \exp(-\frac{y^2t^2}{2})\rho(dy)$. Prove that such a ρ, if it exists, is unique.

METHOD. For the uniqueness of ρ, use Problem II-20. For its existence, use Problems III-1, III-2(4), and IV-20, as well as Paul Lévy's theorem on the convergence of distributions.

SOLUTION. *Uniqueness.* Let $\tilde{\rho}$ be the image of ρ under the mapping $y \mapsto \frac{y^2}{2}$. Then, for $s > 0$, $\widehat{\mu}(\sqrt{s}) = \int_0^\infty \exp(-\frac{y^2}{2}s)\rho(dy) = \int_0^\infty \exp(-xs)\tilde{\rho}(dx)$, and $\widehat{\mu}(\sqrt{s})$ is the Laplace transform of $\tilde{\rho}$. Problem II-20 implies that $\tilde{\rho}$ is unique, and therefore so is ρ.

Existence.
($\Leftarrow$) Let $Y, X_1, \ldots, X_n, \ldots$ be independent real variables such that Y has distribution ρ and X_n has distribution $\nu(dx) = \exp(-\frac{x^2}{2})\frac{dx}{\sqrt{2\pi}}$ for every n. If $a_1^2 + \cdots + a_n^2 = 1$, the

$$(*) \qquad \mathbf{E}[\exp(itY(a_1X_1 + \cdots + a_nX_n))] = \int_0^\infty \exp(-\frac{y^2t^2}{2})\rho(dy).$$

The best way to prove this kind of equality is to use the conditional expectation. Setting $f(z) = \exp(itz)$ and $X = a_1X_1 + \cdots + a_nX_n$ gives

$$\mathbf{E}[f(XY)] = \mathbf{E}[\mathbf{E}[\frac{f(XY)}{Y}]] = \int_0^\infty \mathbf{E}[f(Xy)]\rho(dy).$$

(See Problem IV-34.) $\mathbf{E}[f(Xy)]$ is easy to compute, and $(*)$ is proved.

($\Rightarrow$) We prove this first for the case $\mu(\{0\}) = 0$. Then, for every n, the corresponding μ_n satisfy $\mu_n(\{0\}) = 0$. By Problems III-1 and III-2, if Z_n is a random variable with values in $\mathbf{R}^n$ and distribution μ_n, then $\frac{Z_n}{\|Z_n\|}$ is independent of $\|Z_n\|$ and has distribution σ_{n-1}, the uniform probability measure on the unit sphere S_{n-1} of $\mathbf{R}^n$. Let ν_n be the distribution on $\mathbf{R}$ of $\frac{\sqrt{n}\langle a, Z_n\rangle}{\|Z_n\|}$, where a has norm 1 in $\mathbf{R}^n$. We saw in Problem IV-20 that ν_n converges narrowly to $\nu(dx) = \exp(-\frac{x^2}{\sqrt{2\pi}})$ as $n \to \infty$. Set

$$\alpha_n(t) = \int_{\mathbf{R}^n} \left(\frac{\|z\|}{\sqrt{n}}\right)^{it} \mu_n(dz) \quad \text{and} \quad \beta_n(t) = \int_{\mathbf{R}} |x|^{it}\nu_n(dx).$$

Then, using the independence of $\|Z_n\|$ and $\frac{Z_n}{\|Z_n\|}$,

$$\alpha_n(t)\beta_n(t) = \int_{\mathbf{R}} |z|^{it}\mu(dz) \quad \forall t \in \mathbf{R}.$$

Since $\|x\|^{it}$ is bounded on $\mathbf{R}$ and $\nu_n \to \nu$ narrowly as $n \to \infty$,

$$\lim_{n\to\infty} \beta_n(t) = \int_{\mathbf{R}} |x|^{it} \exp(-\frac{x^2}{2})\frac{dx}{\sqrt{2\pi}} = \frac{2^{it}}{\sqrt{\pi}}\Gamma(\frac{1+it}{2}).$$

The right-hand side is a continuous function of t which is equal to 1 if $t = 0$, has modulus $\frac{1}{\sqrt{\cosh \frac{\pi t}{2}}}$, and never vanishes. Hence

$$\lim_{n\to\infty} \alpha_n(t) = \frac{2^{-it}\sqrt{\pi}}{\Gamma(\frac{1+it}{2})} \int_{\mathbf{R}} |x|^{it} \mu(dx)$$

exists for all t, and the convergence is uniform on compact sets. By Paul Lévy's theorem (IV-4.1.2), this implies that there exists a probability measure ρ on $(0, +\infty)$ such that

$$\lim_{n\to\infty} \alpha_n(t) = \int_0^\infty y^{it} \rho(dy).$$

Therefore $\int_{\mathbf{R}} |z|^{it}\mu(dz) = \int_0^\infty y^{it}\rho(dy) \int_{\mathbf{R}} |x|^{it} \exp(-\frac{x^2}{2})\frac{dx}{\sqrt{2\pi}}$. Since μ is a symmetric probability measure (as the projection of symmetric probability measures), this last equation implies the desired result. For suppose that Y has distribution ρ and X has distribution ν, with X and Y independent. Then the equation implies that the distribution of $\log Y = \log |X|$ is the distribution of $\log |Z|$, where Z has distribution μ, and hence, using symmetry, that the distribution of XY is the distribution of Z. Conditioning with respect to Y,

$$\begin{aligned}
\int_{-\infty}^{+\infty} \exp(itz)\mu(dz) &= \mathbf{E}[\exp(itz)] = \mathbf{E}[\exp(itXY)] \\
&= \mathbf{E}[\mathbf{E}[\exp(itXY)|Y]] \\
&= \mathbf{E}[\exp(-\tfrac{t^2Y^2}{2})] = \int_0^\infty \exp(-\frac{y^2t^2}{2})\rho(dy).
\end{aligned}$$

It remains to consider the case $m = \mu(\{0\}) > 0$. Defining the probability measure μ_1 by $\mu = m\delta_0 + (1-m)\mu_1$, where δ_0 is the Dirac measure at 0, we have $\mu_1 \in \cap_{n=1}^\infty \mathcal{S}_n$, and there exists a probability measure ρ_1 on $(0, +\infty)$ such that $\widehat{\mu}_1(t) = \int_0^\infty \exp(-\frac{y^2t^2}{2})\rho_1(dy)$. Setting $\rho = m\delta_0 + (1-m)\rho_1$ gives the desired result.

REMARK. This property is due to I. Schoenberg (1937).

Problem IV-22. *Let $(X_0, X_1, \ldots, X_d)$ be an $\mathbf{R}^{d+1}$-valued random variable that is radial, i.e. whose distribution is invariant under the group O_{d+1} of $d\times d$ orthogonal matrices. Let $t = (t_1, t_2, \ldots, t_d)$ and let $\|t\| = [t_1^2 + \cdots + t_d^2]^{\frac{1}{2}}$. Prove that $\mathbf{E}[\exp(i\sum_{j=1}^d t_jX_j - \|t\|X_0)] = 1$ for every t in $\mathbf{R}^d$ such that $\mathbf{E}[\exp(-\|t\|X_0)] < \infty$.*

METHOD. Prove the assertion first for $d = 1$ and μ concentrated on the unit circle.

SOLUTION. We show first that

$$\frac{1}{2\pi}\int_0^{2\pi} \exp(it\cos\theta - |t|\sin\theta)d\theta = 1 \quad \forall t \in \mathbf{R}.$$

By symmetry, it suffices to prove this for $t > 0$. In this case,

$$\exp(it\cos\theta) - |t|\sin\theta) = \exp(ite^{i\theta}) = \sum_{n=0}^{\infty} \frac{1}{n!} i^n t^n e^{in\theta},$$

and the result follows since $\int_0^{2\pi} \exp(in\theta)d\theta = 0$ for $n > 0$.

In the general case, note that if

$$\psi(t,s) = \mathbf{E}(\exp(i\sum_{j=1}^{d} t_j X_j - sX_0),$$

where $s \in S = \{s \geq 0 : \mathbf{E}(\exp(-sX_0)) < \infty\}$, then there exists $\varphi : [0,+\infty) \times S \to \mathbf{R}$ such that $\psi(t,s) = \varphi(\|t\|, s)$. (This can be seen as in Problem III-1, by considering the subgroup of O_{d+1} which fixes every point of the axis $(x_0, 0, \ldots, 0)$.) We can then take $t = (t_1, 0, \ldots, 0)$ in $\mathbf{R}^d$: it suffices to show that $\mathbf{E}[\exp(it_1X_1 - |t_1|X_0)] = 1$ for every t_1 such that $|t_1| \in S$. But (X_0, X_1) has a radial distribution in $\mathbf{R}^2$. Let ν be the distribution of $\sqrt{X_0^2 + X_1^2}$; passing to polar coordinates (and omitting some details if $\nu(\{0\}) > 0$), we can write

$$\mathbf{E}[\exp(it_1X_1 - |t_1|X_0)] = \int_0^{\infty} \nu(d\rho) \frac{1}{2\pi} \int_0^{\infty} \exp(i\rho t_1 \cos\theta - \rho|t_1|\sin\theta)d\theta = 1.$$

Problem IV-23. *Let $\{(V_n, W_n)\}_{n=1}^{\infty}$ be a sequence of independent random variables with the same distribution, with values in $\mathbf{R} \times \mathbf{R}^d$ (where $\mathbf{R}^d$ has the Euclidean structure), and satisfying $\mathbf{E}[\log|V_1|] < 0$ and $\mathbf{E}[\log^+ \|W_1\|] < \infty$.*

(1) Prove that $\sum_{n=0}^{\infty} |V_1 \ldots V_n| \; \|W_{n+1}\|$ converges almost surely.

METHOD. Use the Borel-Cantelli lemma to show that $\limsup_{n\to\infty} \|W_n\|^{\frac{1}{n}} \leq 1$; then use the strong law of large numbers. (See Problem IV-7.)

(2) Let μ be the distribution of the $\mathbf{R}^d$-valued random variable that is equal to the sum of the series $\sum_{n=0}^{\infty} V_1 \ldots V_n W_{n+1}$. Show that if ν is a distribution on $\mathbf{R}^d$ whose Fourier transform $\widehat{\nu}$ satisfies

$$\widehat{\nu}(t) = \mathbf{E}[\widehat{\nu}(V_1 t)\exp(i\langle W_1, t\rangle)] \quad \textit{for every } t \textit{ in } \mathbf{R}^d,$$

then $\mu = \nu$.

(3) Let $\{U_n\}_{n=0}^{\infty}$ be a sequence of independent $\mathbf{R}^{d+1}$-valued random variables with the same distribution, the uniform distribution on the unit sphere S_d of $\mathbf{R}^{d+1}$. Let $V_n - 1$ and W_n be the projections of U_n onto $(\mathbf{R}, 0, 0, \dots)$ and onto its orthogonal complement. Prove that if μ is the distribution of $\sum_{n=0}^{\infty} V_1 \dots V_n W_{n+1}$, then $\widehat{\mu}(t) = \exp(-\|t\|)$.
METHOD. Use (2) and Problem IV-22.

(4) Let $\{X_n\}_{n=1}^{\infty}$ be a sequence of independent random variables with values in $\mathbf{N} = \{0, 1, 2, \dots\}$ and with the same distribution, such that X_1 satisfies $p_k = P[X_1 = k] < 1$ for every k in $\mathbf{N}$. Set $q_k = P[X_1 < k]$. Show that if μ is the distribution of $\sum_{n=0}^{\infty} p_{X_1} p_{X_2} \cdots p_{X_n} q_{X_{n+1}}$, then μ is Lebesgue measure on $[0, 1]$.

SOLUTION. (1) $\int_0^\infty P[\log \|W_1\| \geq x]dx = \mathbf{E}[\log \|W_1\|] < \infty$
(See, for example, Problem I-6(2) for $\alpha = 1$.) Hence, for every $\epsilon > 0$,

$$\sum_{n=0}^{\infty} P[\log \|W_n\| \geq k\epsilon] \leq \sum_{n=0}^{\infty} \frac{1}{\epsilon} \int_{n\epsilon}^{(n+1)\epsilon} P[\log \|W_1\| \geq x]dx < \infty.$$

The Borel-Cantelli lemma (I-5.2.8) then implies that $P[\frac{1}{n} \log \|W_n\| \geq \epsilon$ for infinitely many n in $\mathbf{N} \setminus \{0\}] = 0$ for every $\epsilon > 0$. Hence

$$P[\limsup_{n\to\infty} \frac{1}{n} \log \|W_n\| \leq 0] = 1.$$

Next, the law of large numbers implies that

$$P\left[\lim_{n\to\infty} \frac{1}{n} \sum_{k=1}^{n} \log |V_k|\right] = 1,$$

and therefore $P[\limsup_{n\to\infty} \sqrt[n]{|V_1 \dots V_n| \, \|W_{n+1}\|} < 1] = 1$. By Cauchy's criterion, this implies that the series converges almost surely.
(2) Suppose that ν satisfies the given equation. Then it is easy to see by induction on n that

$$\widehat{\nu}(t) = \mathbf{E}\left[\widehat{\nu}(V_1 V_2 \dots V_n t) \exp\left(i\langle t_1, \sum_{k=0}^{n} V_1 \dots V_k W_{k+1}\rangle\right)\right]$$

for every t in $\mathbf{R}^d$ and every positive integer n.

By (1), $V_1 V_2 \dots V_n \to 0$ almost surely as $n \to \infty$. Since $\widehat{\nu}$ is bounded by 1, we can let $n \to \infty$ and apply the dominated convergence theorem to obtain

$$\widehat{\nu}(t) = \mathbf{E}\left[\exp\left(i\langle t, \sum_{k=0}^{\infty} V_1 \dots V_k W_{k+1}\rangle\right)\right] = \widehat{\mu}(t).$$

(3) $\|W_1\| \le \|U_1\| = 1$ implies that $0 = \mathbf{E}[\log^+ \|W_1\|]$. To see that $\mathbf{E}[\log |V_1|] < 0$, observe that V_1 is not concentrated at a single point and that log is a strictly concave function. Hence, by Jensen's inequality (IV-2.5.2), $\mathbf{E}[\log |V_1|] < \log \mathbf{E}[|V_1|]$. Since $-1 < V_1 - 1 < 1$ and $V_1 - 1$ is a symmetric random variable, it follows that $\mathbf{E}[|V_1|] = 1$ and hence that $\mathbf{E}[\log |V_1|] < 0$. Since (2) can be applied, it suffices to check that $\widehat{\nu}(t) = \exp(-\|t\|)$ satisfies

$$\widehat{\nu}(t) = \mathbf{E}[\widehat{\nu}(V_1 t) \exp(i\langle W_1, t\rangle].$$

To do this, we apply Problem IV-22 to U_1 and obtain

$$1 = \mathbf{E}[\exp(i\langle W_1, t\rangle - \|t\|(V_1 - 1))],$$

which implies the desired equality.
(4) We proceed as in (3), applying (2) to $V_n = p_{X_n}$ and $W_n = q_{X_n}$. In this case, if $\widehat{\nu}(t) = \int_0^1 \exp(itx)dx = \frac{1}{t}(\exp(it) - 1)$, then

$$\mathbf{E}[\widehat{\nu}(p_{X_1} t) \exp(iq_{X_1} t)] = \sum_{n=0}^{\infty} \frac{e^{ip_n t} - 1}{p_n t} e^{iq_n t} p_n = \frac{1}{t} \sum_{n=0}^{\infty} (e^{iq_{n+1} t} - e^{iq_n t}) = \widehat{\nu}(t).$$

Problem IV-24. *Let X and Y be independent random variables with the same distribution and with values in Euclidean space $\mathbf{R}^d$, $d > 1$, which satisfy the following conditions: (i) $P[X = 0] = 0$; (ii) $\frac{X}{\|X\|}$ and $\|X\|$ are independent; and (iii) $\frac{X}{\|X\|}$ is uniformly distributed on the sphere S_{d-1}. (That is, the distribution of X is "radial" — see Problem III-2(4).) Prove that*

$$P[\|2X - Y\| \le \|Y\|] < \frac{1}{4}$$

and that this inequality is the best possible.

METHOD. Consider $R = \frac{\|X\|}{\|Y\|}$, use the fact that R and R^{-1} have the same distribution on $(0, +\infty)$, and prove the inequality by first conditioning with respect to $|\log R|$. For the second part, take $\|X\|$ with density $\frac{1}{n} x^{\frac{1-n}{n}}$ on $(0, 1]$ and show that the distribution ν_n of $\exp(-|\log R|)$ tends vaguely to the Dirac measure at 0.

SOLUTION. Set $\theta_1 = \frac{X}{\|X\|}$ and $\theta_2 = \frac{Y}{\|Y\|}$. Then

$$P[\|2X - Y\| \le \|Y\|] = P[\|2R\theta_1 - \theta_2\| \le 1].$$

Since the four random variables θ_1, $\|X\|$, θ_2, and $\|Y\|$ are independent, the same holds for θ_1, θ_2 and R. Moreover, if

$$A = \exp(-|\log R|) = \min\{R, R^{-1}\},$$

then $R = A^\epsilon$, where $P[\epsilon = 1] = P[\epsilon = -1]$ and ϵ is independent of A since R and R^{-1} have the same distribution. If ν denotes the distribution of A, then

$$\begin{aligned} P[\|2R\theta_1 - \theta_2\| \le 1] &= \mathbf{E}[P[\|2R\theta_1 - \theta_2\| \le 1|A]] \\ &= \int_0^1 P[\|2a^\epsilon\theta_1 - \theta_2\| \le 1]\nu(da) \\ &= I^- + I^0 + I^+, \end{aligned}$$

where

$$\begin{aligned} I^+ &= \tfrac{1}{2}\int_{(0,1)} P[\|2a\theta_1 - \theta_2\| \le 1]\nu(da), \\ I^- &= \tfrac{1}{2}\int_{(0,1)} P[\|2a^{-1}\theta_1 - \theta_2\| \le 1]\nu(da), \quad \text{and} \\ I^0 &= P[\|2\theta_1 - \theta_2\| \le 1]\nu(\{1\}). \end{aligned}$$

Let $B(p,1)$ denote the closed ball of radius 1 and center p. If $a < 1$, then $B(0,1)$ and $B(2a^{-1}\theta,1)$ are disjoint; this implies that $P[\|2^{-1}\theta_1 - \theta_2\| \le 1] = P[\theta_2 \in B(2a\theta_1,1)] = 0$ and hence that $I^- = 0$. Similarly, if $a = 1$ then the intersection of $B(0,1)$ and $B(2\theta_1,1)$ is reduced to a point, which implies that $I^0 = 0$ since $d > 1$.

We next show that if $p \neq 0$, then

$$(i) \qquad P[\theta_2 \in B(p,1)] < \frac{1}{2}.$$

To see this, we introduce an orthonormal basis $(e_j)_{j=1}^d$ of $\mathbf{R}^d$ such that $p = \|p\|e_1$. If $(x_j)_{j=1}^d$ are the coordinates of θ_2, then $x_1^2 + \cdots x_d^2 = 1$, and $\theta_2 \in B(p,1)$ if and only if $(x_1 - \|p\|)^2 + x_2^2 + \cdots + x_d^2 \le 1$. This is equivalent to $x_1 \ge \frac{1}{2}\|p\|$, or $\langle\theta_2, p\rangle \ge \frac{1}{2}\|p\|^2$, where $\langle\theta_2, p\rangle$ is the scalar product in $\mathbf{R}^d$. Since the distribution of θ_2 is uniform on S_{d-1},

$$P[\theta_2 \in B(-p,1)] = P[\theta_2 \in B(p,1)].$$

Since the events $\{\langle\theta_2,p\rangle \ge \frac{1}{2}\|p\|^2\}$, $\{\langle\theta_2,-p\rangle \ge \frac{1}{2}\|p\|^2\}$, and $\{|\langle\theta_2,p\rangle| < \frac{1}{2}\|p\|^2\}$ form a partition of the probability space and the third event has positive probability (because $d > 1$),

$$(ii) \qquad 1 = P[|\langle\theta_2,p\rangle| < \frac{1}{2}\|p\|^2] + 2P[\theta_2 \in B(p,1)].$$

This proves (i).

Observe now that, if $a < 1$,

$$(iii) \qquad P[\|2a\theta_1 - \theta_2\| \le 1] = \mathbf{E}[P[\theta \in B(2a\theta_1,1)|\theta_1].$$

By (i), both sides of (iii) are less than $\frac{1}{2}$, which shows that $I^+ < \frac{1}{4}$ and proves the inequality.

To see that this is best possible, note first that, by (ii), $\lim_{p\to 0} P[\theta_2 \in B(p,1)] = \frac{1}{2}$ and hence that

$$\text{(iv)} \qquad \lim_{a\to 0} P[\|2a\theta_1 - \theta_1\| \leq 1] = \frac{1}{2}.$$

We now compute the distribution of R if $\|X\|$ has density $\frac{1}{n}x^{\frac{1-n}{n}}\mathbf{1}_{(0,1]}(x)$. In this case,

$$\mathbf{E}[\exp(it\log R)] = \mathbf{E}[\exp(it\log\|X\|)]\mathbf{E}[\exp(-it\log\|Y\|)] = \varphi_n(t)\varphi_n(-t),$$

where $\varphi_n(t) = \frac{1}{n}\int_0^1 x^{it+\frac{1}{n}-1}dx = \frac{1}{1+int}$. Thus

$$\mathbf{E}[\exp(it\log R] = \frac{1}{1+n^2t^2} = \frac{1}{2(1+int)} + \frac{1}{2(1-int)},$$

which implies that R has density $\frac{1}{2n}x^{\frac{1-n}{n}}$ on $(0,1]$ and density $\frac{1}{2n}x^{-\frac{1+n}{n}}$ on $[1,+\infty)$. Therefore $A = \exp(-|\log R|)$ has the same density ν_n as $\|X\|$. It is clear that ν_n tends vaguely to δ_0 since, for every $c > 0$,

$$\int_0^c \frac{1}{n}x^{\frac{1-n}{n}}dx = c^{\frac{1}{n}} \to 1 \quad \text{as } n \to 0.$$

Let a function f be defined by $f(0) = \frac{1}{2}$ and $f(a) = P[\|2a\theta_1 - \theta_2\| \leq 1]$ for $0 < a \leq 1$. Then f is *continuous* on the *compact set* $[0,1]$, and

$$I_n^+ = \frac{1}{2}\int_0^1 f(a)\nu_n(da) \to \frac{1}{2}\int_0^1 f(a)\delta_0(da) = \frac{1}{4} \quad \text{as } n \to \infty.$$

REMARKS. 1. There is also an explicit expression,

$$\text{(v)} \qquad P[\|2X - Y\| \leq \|Y\|] = \frac{1}{4}\int_0^\infty G(a)d\nu^*(a),$$

where $G(y) = \frac{1}{B(\frac{1}{2},\frac{d-1}{2})}\int_y^\infty \frac{dx}{\sqrt{x}(1+x)^{\frac{d}{2}}}$ and $\nu^*(da)$ is the distribution of $\frac{A^2}{1-A^2}$. This is derived from (iii) by observing that, if u is a unit vector, $P[\theta_2 \in B(2a\theta_1,1)] = P[\langle\theta_2,u\rangle \geq a]$. Since the projection of a uniform distribution on S_{d-1} has the same distribution as $[\frac{X_d^2}{X_1^2+\cdots X_d^2}]^{\frac{1}{2}}$, where the $X_1,\ldots,X_d$ are independent and have distribution $N(0,1)$ (see Problems IV-1 and IV-12), the inequality follows.

2. This inequality is due to A.O. Pittenger, who proves it with the additional hypothesis $P[\|X\| = x] = 0$ for all $x \geq 0$ (1981).

3. Relaxing the hypothesis of the problem to $P[\|X\| = 0] = 0$ easily yields the upper bound

$$P[\|2X - Y\| \leq \|Y\|] < p + (1-p)^2\frac{1}{4},$$

where $p = P[\|X\| = 0] < 1$, and this again is best possible. Note also that $P[\|2X - Y\| < \|Y\|] = \frac{(1-p)^2}{4} < \frac{1}{4}$ in all cases.

Problem IV-25. *Let H be a separable Hilbert space and let p_U denote the orthogonal projection of H onto a subspace U. Define the Boolean algebra $\mathcal{B}$ of subsets B of H for which there exist a finite-dimensional subspace V of H and a Borel set B_V of V such that $B = p_V^{-1}(B_V)$. Let $\sigma(\mathcal{B})$ denote the σ-algebra generated by $\mathcal{B}$.*
(1) Show that $\{x : \|x\| \le r\} \in \sigma(\mathcal{B})$ if $r > 0$.
METHOD. Use the fact that, since H is separable, there exists an increasing sequence $\{V_n\}_{n=1}^{\infty}$ of finite-dimensional subspaces of H such that $\cup_{n=1}^{\infty} V_n$ is dense in H.

(2) A cylindrical probability *on H is given by probabilities μ_V on each finite-dimensional subspace V of H such that, if $V_1 \subset V_2$, the image of μ_{V_2} under p_{V_1} is μ_{V_1}. For $B \in \mathcal{B}$, let E_B denote the set of finite-dimensional subspaces V such that there exists a Borel subset B_V of V with $B = p_V^{-1}(B_V)$. Prove that $V \mapsto \mu_V(B_V)$ is constant on E_B. Denoting this constant by $\mu(B)$, prove that μ is finitely additive on $\mathcal{B}$.*
(3) Consider the cylindrical probability defined as follows. Let ρ be a probability measure on $[0, +\infty)$ and let μ_V be defined by its Fourier transform,

$$\widehat{\mu}_V(t) = \int_V \exp(i\langle x, t\rangle)\mu_V(dx) = \int_0^{\infty} \exp\left(-\frac{y^2\|t\|^2}{2}\right)\rho(dy) \quad \text{for } t \in V.$$

Show that μ is not σ-additive on $\mathcal{B}$ if $\rho(\{0\}) < 1$.

METHOD. Otherwise μ could be extended to a σ-additive probability measure μ on $\sigma(\mathcal{B})$. Use Problems I-10 and IV-6 to show that this would imply $\mu(\{x : \|x\| \le r\}) = \rho(\{0\})$ for $r > 0$.

SOLUTION. (1) We know that

$$\begin{aligned} \|x\| &= \sup\{|\langle y, x\rangle| : \|y\| \le 1\} \\ &= \sup\{|\langle y, x\rangle| : y \in \cup_{n=1}^{\infty} V_n \text{ and } \|y\| \le 1\}. \end{aligned}$$

Hence

$$\begin{aligned} \{x : \|x\| \le r\} &= \{x : |\langle y, x\rangle| \le r \ \ \forall y \in \cup_{n=1}^{\infty} V_n \text{ with } \|y\| \le 1\} \\ &= \cap_{n=1}^{\infty}\{x : |\langle y, x\rangle| \le r \ \ \forall y \in V_n \text{ with } \|y\| \le 1\}. \end{aligned}$$

If $B_{V_n} = \{x \in V_n : |\langle y, x\rangle| \le r \ \ \forall y \in V_n \text{ with } \|y\| \le 1\}$, then clearly B_{V_n} is a Borel subset of V_n. Thus

$$\{x : \|x\| \le r\} = \cap_{n=1}^{\infty} p_{V_n}^{-1}(B_{V_n}),$$

which implies that $\{x : \|x\| \le r\}$ is a countable intersection of elements of $\mathcal{B}$.

(2) A routine verification.
(3) Let $\{V_n\}_{n=1}^{\infty}$ be an increasing sequence of subspaces of H such that V_n has dimension n and $\cup_{n=1}^{\infty} V_n$ is dense in H. Suppose that μ exists on $\sigma(\mathcal{B})$ as defined in (2). Then, since

$$\{x : \|x\| \leq r\} \subset \{x : x \in V_n \text{ and } \|x\| \leq r\}$$

for every n, we can write

$$\mu(\{x : \|x\| \leq r\}) \leq \mu_{V_n}(\{x : x \in V_n \text{ and } \|x\| \leq r\}) = a_n.$$

To show that $a_n \to 0$ as $n \to \infty$, consider the independent real random variables $Y, X_1, X_2, \ldots, X_n, \ldots$ such that Y has distribution ρ and the X_j have distribution $\nu(dx) = \exp(-\frac{x^2}{2})\frac{dx}{\sqrt{2\pi}}$. Then

$$\mathbf{E}[\exp(i\sum_{j=1}^{n} t_j Y X_j] = \int_0^{\infty} \exp\left(-\frac{y^2}{2}\sum_{j=1}^{n} t_j^2\right)\rho(dy) \quad \forall t_j \in \mathbf{R},$$

and hence

$$a_n = P_r[(X_1^2 + \cdots + X_n^2)Y^2 \leq r^2].$$

But, by Problems I-10 and IV-6, for every $\epsilon > 0$

$$P\left[\left|\frac{X_1^2 + \cdots + X_n^2}{n}Y^2 - Y^2\right| \geq \epsilon\right] \to 0 \quad \text{as } n \to \infty.$$

A standard argument then shows that

$$a_n = P\left[\frac{X_1^2 + \cdots + X_n^2}{n}Y^2 \leq \frac{r^2}{n}\right] \to P[Y = 0] \quad \text{as } n \to \infty.$$

It follows that $1 = \mu(\cup_{n=1}^{\infty}\{x : \|x\| \leq n\}) < 1$, a contradiction.

Problem IV-26. *In Euclidean space $\mathbf{R}^n$, consider the positive quadratic form q defined by $q(x) = \sum_{k=1}^{n} \lambda_k x_k^2$, where $x = \{x_k\}_{k=1}^{n}$ and $\lambda_k \geq 0$. Set $\|q\| = \sum_{k=1}^{n} \lambda_k$.*
(1) If X is an $\mathbf{R}^n$-valued random variable such that

$$\mathbf{E}(\exp(i\langle X, t\rangle)) = \exp\left(-\frac{\|t\|^2}{2}\right),$$

show that $P[q(X) \geq r^2] \leq \frac{\|q\|}{R^2}$ for every $r > 0$.
METHOD. Use Chebyshev's inequality, Problem IV-6.

(2) Let μ be a probability measure on $\mathbf{R}^n$ with Fourier transform $\widehat{\mu}(t) = \int_{\mathbf{R}^n} \exp(i\langle x,t\rangle)\mu(dx)$ and let $\epsilon > 0$ be such that $|1-\widehat{\mu}(t)| \leq \epsilon$ for every t in $\mathbf{R}^n$ with $q(t) \leq 1$. Prove that, for every $r > 0$,

$$\int_{\mathbf{R}^n} \exp\left(-\frac{\|x\|^2}{2r^2}\right)\mu(dx) \geq 1 - \epsilon - \frac{2\|q\|}{r^2}.$$

(3) Prove that, for every r, $R > 0$,

$$\mu(\{x : \|x\| \leq R\} \geq 1 - \epsilon - \frac{2\|q\|}{r^2} - \exp\left(-\frac{R^2}{2r^2}\right).$$

Conclude that there exists a number $R(\|q\|,\epsilon)$ such that

$$\mu(\{x : \|x\| \leq R(\|q\|,\epsilon)\} \geq 1 - 2\epsilon.$$

SOLUTION.

(1) $P[\lambda_1 X_1^2 + \cdots + \lambda_n X_n^2 \geq r^2] \leq \frac{1}{r^2}\mathbf{E}(\lambda_1 X_1^2 + \cdots + \lambda_n X_n^2) = \frac{1}{r^2}\sum_{k=1}^{n}\lambda_j.$

(2)

$$\begin{aligned}\int_{\mathbf{R}^n} \exp\left(-\frac{\|x\|^2}{2r^2}\right)\mu(dx) &= \int_{\mathbf{R}^n}\mu(dx)\int_{\mathbf{R}^n}\frac{r^n}{(\sqrt{2\pi})^n}\exp\left(i\langle x,t\rangle - \frac{\|rt\|^2}{r}\right)\\ &= \int_{\mathbf{R}^n}\frac{r^n}{(\sqrt{2\pi})^n}\exp\left(-\frac{\|rt\|^2}{2}\right)\widehat{\mu}(t)dt\end{aligned}$$

by Fubini's theorem. (This is Parseval's identity.) But

$$\begin{aligned}1 - &\int_{\mathbf{R}^n}\frac{r^n}{(\sqrt{2\pi})^n}\exp(-\frac{\|rt\|^2}{2})\widehat{\mu}(t)dt\\ &= \int_{\mathbf{R}^n}\frac{r^n}{(\sqrt{2\pi})^n}\exp(-\frac{\|rt\|^2}{2})(1-\widehat{\mu}(t))dt\\ &\leq \int_{\mathbf{R}^n}\frac{r^n}{(\sqrt{2\pi})^n}\exp(-\frac{\|rt\|^2}{2})|1-\widehat{\mu}(t)|dt\\ &= \int_{\{t:q(t)\leq 1\}} + \int_{\{t:q(t)\geq 1\}}\\ &\leq \epsilon + 2\int_{\{t:q(t)\geq 1\}}\frac{r^n}{(\sqrt{2\pi})^n}\exp(-\frac{\|rt\|^2}{2}).\end{aligned}$$

The last integral is that of (1), by the change of variable $x = rt$; this completes the proof.

(3)

$$\begin{aligned}\int_{\mathbf{R}^n}\exp\left(-\frac{\|x\|^2}{2r^2}\right)\mu(dx) &= \int_{\{x:\|x\|\leq R\}} + \int_{\{x:\|x\|>R\}}\\ &\leq \mu(\{x : \|x\| \leq R\}) + \mathrm{e}^{-\frac{R^2}{2r^2}}.\end{aligned}$$

This, with (2), proves the first inequality. Next, consider $\varphi(s) = 2\|q\|s + e^{-\frac{R^2 s}{2}}$ for $s > 0$. Its minimum, attained at $s_0 = \frac{2}{R^2}\log\frac{4\|q\|}{R^2}$, is $\varphi(s_0) = \frac{4\|q\|}{R^2}\log\frac{R^2 e}{4\|q\|}$. Since $\mu(\{x : \|x\| \leq R\}) \geq 1 - \epsilon - \varphi(\frac{1}{r^2})$ for every r, we obtain

$$\mu(\{x : \|x\| \leq R\}) \geq 1 - \epsilon - \varphi(s_0).$$

Taking R sufficiently large that $\varphi(s_0) \leq \epsilon$ gives the desired result.

REMARK. This result is called Minlos's lemma (1959).

Problem IV-27. *The notation is that of Problem IV-25 and $\mu = (\mu_V)_V$ is a cylindrical probability on H. A positive quadratic form q on H is a bounded linear mapping $A : H \to H$ such that $q(x) = \langle Ax, x\rangle \geq 0$ for every x. If the dimension of V is n, there exist a basis $b = \{b_1, \ldots, b_n\}$ of V and nonnegative numbers $\lambda_1, \ldots, \lambda_n$ such that if $\sum_{k=1}^n x_k b_k$ is in V, then $q(x) = \sum_{k=1}^n \lambda_k x_k^2$. Moreover, the distribution of the $\{\lambda_k\}_{k=1}^n$ is independent of b, and we may set $\|q_V\| = \sum_{k=1}^n \lambda_k$. This implies that $\|q_{V_1}\| \leq \|q_{V_2}\|$ if $V_1 \subset V_2$, and we set $\|q\| = \sup_V \|q_V\| \leq +\infty$.*
(1) Let $\widehat{\mu}_V(t) = \int_V \exp(i\langle x, t\rangle)\mu_V(dx)$ for $t \in V$. Show that $\widehat{\mu}_{V_1}(t) = \widehat{\mu}_{V_2}(t)$ if $t \in V_1 \cap V_2$.
(2) Set $\widehat{\mu}(t) = \widehat{\mu}_V(t)$ if $t \in V$. Suppose that, for all $\epsilon > 0$, there exists a positive quadratic form q_ϵ on H such that $\|q_\epsilon\| < \infty$ and $|1 - \widehat{\mu}(t)| \leq \epsilon$ for all t such that $q_\epsilon(t) \leq 1$. Deduce from Problem IV-26 that, for all $\epsilon > 0$, there exists $R(\epsilon)$ such that

$$\mu_V(\{x : x \in V \text{ and } \|x\| \leq R(\epsilon)\}) \geq 1 - 2\epsilon \quad \text{for every } V.$$

(3) With the preceding hypotheses, prove that μ is a σ-additive probability measure on the Boolean algebra $\mathcal{B}$ by showing that if $A_n \in \mathcal{B}$, $A_n \supset A_{n+1}$, and $\mu(A_n) \geq \delta > 0$ for every n, then $\cap_{n=1}^\infty A_n \neq \emptyset$.

METHOD. Let V_n be a finite-dimensional subspace of H containing a Borel set A'_n such that $A_n = p_{V_n}^{-1}(A'_n)$ and let $B'_n(R)$ be the closed ball of radius R in V_n. We may assume that $V_n \subset V_{n+1}$. Construct compact sets K'_n of V_n contained in $A'_n \cap B'_n(R)$, introduce $K_n = p_{V_n}^{-1}(K'_n)$, and use the fact that $C_n = K_n \cap \ldots \cap K_n \cap \{x : \|x\| \leq R\}$ is a decreasing sequence of compact sets in the weak topology on H.

SOLUTION. (1) and (2) are immediate.
(3) With $R(\epsilon)$ defined as in (2), set $R = R(\frac{\delta}{10})$. Then, if $B_n(R) = p_{V_N}^{-1}(B'_n(R))$,

$$\mu(A_n^c \cup B_n(R)^c) \leq \mu(A_n^c) + \mu(B_n(R)^c) \leq 1 - \delta + 2\frac{\delta}{10} = 1 - \frac{4\delta}{5},$$

and hence $\mu(A_n \cap B_n(R)) \geq \frac{4\delta}{5}$. Let K'_n be a compact subset of V_n such that $K'_n \subset A'_n \cap B_n(R)$ and $\mu(A_n \cap B_n(R) \setminus K_n) \leq \frac{\delta}{5 \cdot 2^n}$. Then $\mu(K_n) \geq \frac{4\delta}{5} - \frac{\delta}{5 \cdot 2^n}$ and

$$\begin{aligned}
\mu(K_1 \cap \ldots \cap K_n) &= \mu(A_n \cap B_n(R)) - \mu(A_N \cap B_n(R) \setminus (K_1 \cap \ldots \cap K_n)) \\
&\geq \mu(A_n \cap B_n(R)) - \sum_{k=1}^{n} \frac{\delta}{5 \cdot 2^k} = \frac{4\delta}{5} - \frac{\delta}{5} = \frac{3\delta}{5}.
\end{aligned}$$

Therefore $\mu(K_1 \cap \ldots \cap B_m(R)) \geq \frac{2\delta}{5}$ for all m, and hence $K_1 \cap \ldots \cap K_n \cap B_n(R)$ is nonempty. It follows that C_n is nonempty. But the K_n are closed subsets of H in the norm topology, and hence in the weak topology on H. It is also known (Banach-Alaoglu theorem) that the closed ball of radius $r > 0$ in H is closed in the weak topology. Hence C_n is compact in this weak topology. A classical theorem of general topology states that the intersection of a decreasing family of nonempty compact sets is nonempty. Thus

$$\emptyset \neq \cap_{n=1}^{\infty} C_n \subset \cap_{n=1}^{\infty} A_n.$$

REMARK. This result is due to Minlos (1959).

Problem IV-28. *Let $\{X_n\}_{n \geq 1}$ be a sequence of independent random variables with the same distribution defined by $P[X_n = 1] = P[X_n = -1] = \frac{1}{2}$. Compute the limiting distribution as $n \to \infty$ of*

$$Y_n = [1 + 4 + 9 + \ldots + n^2]^{-\frac{1}{2}} [X_1 + 2X_2 + 3X_3 + \ldots + nX_n].$$

METHOD. Consider the characteristic function of Y_n.

SOLUTION. Set $\sigma_n = [1 + 4 + 9 + \cdots + n^2]^{\frac{1}{2}}$. Then, for t real,

$$\mathbf{E}(\exp(itY_n)) = \cos \frac{t}{\sigma_n} \cos \frac{2t}{\sigma_n} \ldots \cos \frac{nt}{\sigma_n}.$$

If $n \to +\infty$, then $\sigma_n n^{-\frac{3}{2}} \to \frac{1}{\sqrt{3}}$. This can be seen in several ways, for example by bracketing the partial sums of the divergent series $\sum_{k=1}^{\infty} k^2$ by integrals

$$\frac{n^3}{3} = \int_0^n x^2 dx \leq \sum_{k=1}^{n} k^2 = \sigma_n^2 \leq \int_1^{n+1} x^2 dx = \frac{n^3 + 3n^2 + 3n}{3},$$

or by using the classical formula

$$\sigma_n^2 = \frac{n(n+1)(2n+1)}{6}.$$

Thus $\frac{n}{\sigma_n} \to 0$ as $n \to \infty$, and for every $t \in \mathbf{R}$ there exists N such that $n \geq N$ implies $\sup_{k \leq n} \frac{k|t|}{\sigma_n} = \frac{n|t|}{\sigma_n} < \frac{\pi}{2}$. If $n \geq N$, we can therefore write

$$\log \mathbf{E}[\exp(itY_n)] = \sum_{k=1}^{n} \log \cos \frac{kt}{\sigma_n} = -\frac{t^2}{2} + \sum_{k=1}^{n} \frac{k^2t^2}{\sigma_n^2} \epsilon(\frac{kt}{\sigma_n}).$$

Using the limiting expression for the Taylor series of $\log \cos x$ in a neighborhood of 0, we obtain

$$\log \cos x = -\frac{x^2}{2} + x^2 \epsilon(x), \quad \text{where } \epsilon(x) \to 0 \text{ as } x \to 0.$$

Hence, for fixed t, the sequence $c_n = \sup_{k \leq n} |\epsilon(\frac{kt}{\sigma_n})|$, with $n \geq N$, tends to 0 as $n \to \infty$, and

$$\left| \sum_{k=1}^{n} \frac{k^2t^2}{\sigma_n^2} \epsilon \left(\frac{kt}{\sigma_n} \right) \right| \leq c_n t^2 \to 0 \quad \text{as } n \to \infty.$$

Therefore $\mathbf{E}[\exp(itY_n)] \to \mathrm{e}^{-\frac{t^2}{2}}$.

The distribution of Y_n thus tends to the probability measure with density $\frac{1}{\sqrt{2\pi}} \exp(-\frac{x^2}{2})$, by Paul Lévy's theorem (IV-4.1.2) and the known result on characteristic functions of Gaussian distributions (see IV-4.3.2, Lemma (ii)).

REMARK. This is a simple special case of Lindeberg's theorem, which is a significant generalization of Laplace's theorem, IV-4.3.1 (also often called the central limit theorem). Lindeberg's theorem is stated as follows: If (i) the real random variables $\{X_n\}_{n=1}^{\infty}$ are independent (but do not necessarily have the same distribution); (ii) for every n, $\mathbf{E}(X_n) = 0$ and $\sigma_n^2 = \mathbf{E}[(X_1 + \cdots + X_n)^2] < \infty$; and (iii) for every ϵ,

$$\mathbf{E}\left[\sum_{k=1}^{n} f_\epsilon(\frac{X_k}{\sigma_n}) \right] \to 0, \quad \text{where } f_\epsilon(x) = x^2 \mathbf{1}_{[\epsilon, +\infty)}(x),$$

then the distribution of $\frac{1}{\sigma_n}(X_1 + \cdots + X_n)$ tends to the Gaussian distribution $N(0,1)$ as above.

Problem IV-29. *Consider the Gaussian distribution $\mu(dx) = (2\pi)^{-1} \exp(\frac{x^2}{2})dx$ on the real line and the Hilbert space $L^2(\mu)$ of functions that are square integrable with respect to μ, with the scalar product*

$$\langle f, g \rangle = \int_{-\infty}^{+\infty} f(x)\overline{g(x)}\mu(dx).$$

The Hermite polynomials $\{H_n(x)\}_{n=0}^{\infty}$ *are defined by*

$$\sum_{n=0}^{\infty} H_n(x)(it)^n = \exp(itx + \frac{t^2}{2}) = \varphi(t,x) \quad \forall t \in \mathbf{C}.$$

Assume without proof that this implies

$$(*) \qquad \sum_{n=0}^{\infty} |H_n(x)|\ |t|^n \leq \exp\left(|t|\ |x| + \frac{|t|^2}{2}\right).$$

(1) By computing $\langle\varphi(t,.),\varphi(s,.)\rangle$ *in two different ways, show that* $\langle H_n, H_m\rangle = 0$ *if* $n \neq m$ *and that* $\langle H_n, H_n\rangle = \frac{1}{n!}$. *Using the uniqueness of the Fourier transform, show that if* f *in* $L^2(\mu)$ *satisfies* $\langle f, H_n\rangle = 0$ *for every* n*, then* $f = 0$.
(2) Show that $H'_{n-1}(x) = H_n(x)$ *and that* $(n+1)H_{n+1}(x) = xH_n(x) - H_{n-1}(x)$ *if* $n \geq 1$.
(3) Let $f \in L^2(\mu)$ *and let* $f_n = n!\langle f, H_n\rangle$. *Show that* $f = \sum_{n=0}^{\infty} f_n H_n$ *(where the convergence of the series is in the* $L^2(\mu)$ *sense). If, moreover,* f' *exists (in the sense that* $F(x) = f(0) + \int_0^x f'(t)dt$ *for every* x*) and belongs to* $L^2(\mu)$*, show that* $f' = \sum_{n=0}^{\infty} f_{n+1}H_n$.
METHOD. Compute $\langle f', H_n\rangle$ by means of an integration by parts and (2).

(4) Prove H. Chernoff's inequality: If X *is a Gaussian random variable with distribution* μ *and if* f *is a real-valued function such that both* $f'\mathbf{E}[|f'(X)|^2]$ *and* $\mathbf{E}[|f(X)|^2]$ *exist, then* $\mathbf{E}[|f'(X)|^2] \geq$ *variance of* $f(X)$. *Analyze the case of equality.*

SOLUTION. (1) By definition, $\langle\varphi(t,.),\varphi(s,.)\rangle = \exp(\frac{t^2}{2} + \frac{s^2}{2}) \int_{-\infty}^{+\infty} \mathrm{e}^{ix(t+1)}$ $\mu(dx)$, and this equals $\exp(-ts)$ by computing the Fourier transform of μ. By $(*)$ and the dominated convergence theorem, we can interchange the integration and the double summation to obtain

$$\langle\varphi(t,.),\varphi(s,.)\rangle = \sum_{n=0}^{\infty}\sum_{n=0}^{\infty}\langle H_n, H_m\rangle(it)^n(is)^m = \exp(-ts) = \sum_{n=0}^{\infty}\frac{(it)^n(is)^n)}{n!},$$

which gives the orthogonality and the norm of the Hermite polynomials. To see that they form a complete system, let $f \in L^2(\mu)$ such that $\langle f, H_n\rangle = 0$ for every n. Then, using $(*)$ again, we have $\langle f, \varphi(t,.)\rangle = 0$ for all t in $\mathbf{C}$, and hence

$$\int_{-\infty}^{+\infty} f(x)\mathrm{e}^{-\frac{x^2}{2}}\mathrm{e}^{ixt}dx = 0 \quad \text{for every } t.$$

Since $f \in L^2(\mu)$, this implies that $F(x)\mathrm{e}^{-\frac{x^2}{2}}$ is integrable and has Fourier transform zero, hence that $f = 0$.

(2) In general, if $h(t) = \sum_{n=0}^{\infty} h_n(it)^n$ is the sum of a power series with radius of convergence $R > 0$ and $e^{ixt}h(t) = \sum_{n=0}^{\infty} h_n(x)(it)^n$, the polynomials $(h_n(x))_{n\geq 0}$ are given by

$$h_n(x) = \sum_{k=0}^{n} h_{n-k}\frac{x^k}{k!} \qquad \text{(Cauchy product of two series)}.$$

Thus

$$h'_n(x) = \sum_{k=1}^{n} h_{n-k}\frac{x^{k-1}}{(k-1)!} = \sum_{k=0}^{n-1} h_{n-1-k}\frac{x^k}{k!} \quad \text{for } n \geq 1$$

is also a Cauchy product of two series, and

$$\sum_{n=0}^{\infty} h'_n(x)(it)^n = e^{itx}\sum_{n=1}^{\infty} h_{n-1}(it)^n$$

has the same radius of convergence R as $\sum_{n=1}^{\infty} h_{n-1}(it)^n$. It follows that $\sum_{n=0}^{\infty} h'_n(x)(it)^n = e^{itx}ith(t)$ and hence that $h'_n(x) = h_{n-1}(x)$ for $n \geq 1$. Applying this to $h(t) = \exp(\frac{t^2}{2})$ and $R = \infty$ gives $H'_{n-1}(x) = H_n(x)$.

Finally, differentiating the equation defining the H_n with respect to t, we easily obtain

$$(n+1)H_{n+1}(x) = xH_n(x) - H_{n-1}(x).$$

(3) $\sum_{n=0}^{\infty} f_n H_n$ converges in $L^2(\mu)$ since, by Bessel's inequality,

$$\sum_{k=0}^{\infty} \frac{|f_k|^2}{k!} = \sum_{k=0}^{\infty} k!|\langle f, H_k\rangle|^2 \leq \langle f, f\rangle.$$

Since $\langle f - \sum_{n=0}^{\infty} f_n H_n, H_k\rangle = 0$ for every k, it follows from (1) that $f = \sum_{n=0}^{\infty} f_n H_n$.

We now compute $\langle f', H_n\rangle = \lim_{T\to\infty} \int_{-T}^{+T} f'(x)H_n(x)\mu(dx)$.

$$\int_{-T}^{+T} f'(x)H_n(x)e^{-\frac{x^2}{2}}\frac{dx}{\sqrt{2\pi}} = \left[f(x)H_n(x)\frac{e^{-\frac{x^2}{2}}}{\sqrt{2\pi}}\right]_{-T}^{+T} - \int_{-T}^{+T} f(x)(H'_n(x) - xH_n(x))e^{-\frac{x^2}{2}}\frac{\sqrt{2\pi}}{}.$$

The integrated part tends to 0 as $T \to +\infty$, since $g_n(x) = f(x)H_n(x)e^{-\frac{x^2}{2}}$ has an integrable derivative and therefore tends to limits as $x \to \pm\infty$. Moreover, g_n is integrable; thus $\lim_{x\to\pm\infty} g_n(x) = 0$. As for the non-integrated part, its limit as $T \to +\infty$ is $(n+1)\langle f, H_{n+1}\rangle$ by (2). Hence $n!\langle f', H_n\rangle = (n+1)!\langle f, H_{n+1}\rangle = f_{n+1}$, which gives $f' = \sum_{n=0}^{\infty} f_{n+1}H_n$.

(4) $\mathbf{E}(f'^2(X)) = \sum_{n=0}^{\infty} f_{n+1}^2\langle H_n, H_n\rangle = \sum_{n=0}^{\infty}\frac{f_{n+1}^2}{n!}$ by the preceding computation. $\mathbf{E}[f(X)] = f_0$ and the variance of $f(x)$ is

$$\mathbf{E}((f(X) - f_0)^2) = \sum_{n=1}^{\infty}\frac{f_n^2}{n!} = \sum_{n=0}^{\infty}\frac{f_{n+1}^2}{(n+1)!}.$$

Since $\frac{1}{n!} \geq \frac{1}{(n+1)!}$, this proves H. Chernoff's inequality. Equality holds if and only if

$$\sum_{n=0}^{\infty} f_{n+1}^2\left(\frac{1}{n!} - \frac{1}{(n+1)!}\right) = 0, \quad \text{hence} \quad 0 = f_2 = f_3 = \dots,$$

which implies that $f(x) = f_0 + f_1 x$.

REMARK. The definition of the H_n implies that $(n+1)H_{n+1} = xH_n - H_{n-1}$ for $n \geq 1$, and hence that $(n+1)|H_n| \leq |x|\,|H_n| + |H_{n-1}|$. For $t \geq 0$, set $\psi(t,x) = \sum_{n=0}^{\infty} |H_n| t^n$. Then $\frac{\partial \psi}{\partial t}(t,x) \leq (|x| + t)\psi(t,x)$, $\psi(0,x) = 1$, and $\psi(t,x) > 0$. Thus $\int_0^t \frac{1}{\psi}(\frac{\partial}{\partial t}\psi)(s,x)ds \leq \int_0^t (|x| + s)ds$, which leads to $(*)$.

Problem IV-30. *Let (X,Y) be a Gaussian random variable with values in $\mathbf{R}^2$ such that X and Y have distribution $\mu(dx) = (2\pi)^{-\frac{1}{2}} \exp(-\frac{x^2}{2})dx$.*
(1) For the Hermite polynomials defined in Problem IV-29, prove that

$$H_n(y\cos\theta + z\sin\theta) = \sum_{k=0}^{n} H_k(y)\cos^k\theta\, H_{n-k}(z)\sin^{n-k}\theta.$$

(2) Assume that $\cos\theta = \mathbf{E}(XY) \neq \pm 1$ and define the random variable $Z = \dfrac{X - Y\cos\theta}{\sin\theta}$. Verify that Y and Z are independent and use (1) to prove that $\mathbf{E}[H_n(X)|Y] = H_n(Y)(\mathbf{E}(XY))^n$.
(3) Prove Gebelein's inequality: If $f \in L^2(\mu)$ with $\mathbf{E}(f(X)) = 0$, then

$$\mathbf{E}[(\mathbf{E}[f(X)|Y])^2] \leq (\mathbf{E}(XY))^2 \mathbf{E}(f^2(X)).$$

Analyze the case of equality.

METHOD. Write $f = \sum_{n=1}^{\infty} f_n H_n$ as in Problem IV-29.

SOLUTION.
(1)

$$\begin{aligned}
&\sum_{n=0}^{\infty} H_n(y\cos\theta + z\sin\theta)(it)^n \\
&\quad = \exp(ity\cos\theta + \tfrac{t^2}{2}\cos^2\theta)\exp(itz\sin\theta + \tfrac{t^2}{2}\sin^2\theta) \\
&\quad = \left(\sum_{n=0}^{\infty} H_n(y)\cos^n\theta(it)^n\right)\left(\sum_{m=0}^{\infty} H_m(z)\sin^m\theta(it)^m\right).
\end{aligned}$$

(2) If $\mathbf{E}(XY) = \epsilon$, where $\epsilon = \pm 1$, then $X = \epsilon Y$ since $\mathbf{E}[(X - \epsilon Y)^2] = 0$. It is trivial that $\mathbf{E}[H_n(-Y)|Y] = H_n(Y)(-1)^n$ since $H_n(y)$ is a polynomial which has the same parity as n. If $-1 < \cos\theta = \mathbf{E}(XY) < 1$, then Z is

defined, $\mathbf{E}(Z) = 0$, and $\mathbf{E}(YZ) = 0$. Since (X, Y, Z) is Gaussian in $\mathbf{R}^3$, this implies that Y and Z are independent. Hence

$$\begin{aligned}\mathbf{E}[H_n(X)|y] &= \mathbf{E}[H_n(Y\cos\theta + Z\cos\theta)|Y] \\ &= \textstyle\sum_{k=0}^{n} \mathbf{E}[H_k(Y)H_{n-k}(Z)|Y]\cos^k\theta\sin^{n-k}\theta.\end{aligned}$$

Since Y and Z are independent,

$$\mathbf{E}[H_k(Y)H_{n-k}(Z)|Y] = H_k(Y)\mathbf{E}[H_{n-k}(Z)].$$

But $\mathbf{E}[H_{n-k}(Z)] = \langle H_{n-k}, H_0\rangle = 0$ if $n-k > 0$. Thus $\mathbf{E}[H_n(X)|Y] = H_n(Y)\cos^n\theta$.
(3) Write $f \in L^2(\mu)$ as $f = \sum_{n=0}^{\infty} f_n H_n$. The hypothesis $\mathbf{E}(f(X)) = 0$ implies that $f_0 = 0$. Let $(\Omega, \mathcal{A}, P)$ be the probability space on which the random variables X and Y are defined. Then $f(X) = \sum_{n=1}^{\infty} f_n H_n(X)$, where the sum of the series is relative to $L^2(\Omega, \mathcal{A}, P)$. Since conditional expectation is a projection in $L^2(\Omega, \mathcal{A}, P)$ and therefore a continuous linear operator, we can write

$$\mathbf{E}[f(X)|Y] = \sum_{n=1}^{\infty} f_n \mathbf{E}[H_n(X)|Y] = \sum_{n=1}^{\infty} f_n H_n(Y)(\mathbf{E}(XY))^n.$$

Hence, using orthogonality,

$$\mathbf{E}[(\mathbf{E}[f(X)|Y])^2] = \sum_{n=1}^{\infty} f_n^2 \frac{(\mathbf{E}(XY))^{2n}}{n!} \leq \sum_{n=1}^{\infty} \frac{f_n^2}{n!}(\mathbf{E}(XY))^2,$$

and equality can occur only if

$$f_n^2(\mathbf{E}(XY))^2[1 - (\mathbf{E}(XY))^{n-1}] = 0 \quad \text{for all } n \geq 1.$$

This implies that either $\mathbf{E}(XY) = 0$ (independence), $\mathbf{E}(XY) = 1$ $(X = Y)$, $\mathbf{E}(XY) = -1$ $(X = -Y)$ and f is odd, or $0 < |\mathbf{E}(XY)| < 1$ and $f(x) = f_1 x$.

Problem IV-31. *Let H_n be the nth Hermite polynomial described in Problem IV-29 and compute*

$$\int_{-\infty}^{+\infty} e^{ixt} H_n(x) e^{-\frac{x^2}{2}} \frac{dx}{\sqrt{2\pi}} = \widehat{H_n\mu}(t).$$

Use this to find

$$\int_{-\infty}^{+\infty} e^{ixt} x^n e^{-\frac{x^2}{2}} \frac{dx}{\sqrt{2\pi}}.$$

SOLUTION. $\sum_{n=0}^{\infty} H_n(x)(is)^n = \exp(isx + \frac{s^2}{2})$ by definition. Hence

$$\int_{-\infty}^{+\infty} \exp[i(s+t)x + \frac{s^2}{2} - \frac{x^2}{2}]\frac{dx}{\sqrt{2\pi}} = \sum_{n=0}^{\infty}(is)^n \widehat{H_n\mu}(t) = \exp(-st - \frac{t^2}{2}).$$

$\sum$ and $\int$ can be interchanged by $(*)$ of Problem IV-29, and this gives

$$\sum_{n=0}^{\infty} \widehat{H_n\mu}(t)(is)^n = \sum_{n=0}^{\infty} \mathrm{e}^{-\frac{t^2}{2}} \frac{(it)^n}{n!}(is)^n$$

$$\widehat{H_n\mu}(t) = \mathrm{e}^{-\frac{t^2}{2}} \frac{(it)^n}{n!}.$$

By the Fourier inversion formula,

$$\frac{1}{2\pi}\int_{-\infty}^{+\infty} \mathrm{e}^{-ixt-\frac{t^2}{2}} \frac{(it)^n}{n!} dt = H_n(x)\frac{\mathrm{e}^{-\frac{x^2}{2}}}{\sqrt{2\pi}}.$$

Writing $-t$ for t in the integral, then interchanging the roles of x and t, we obtain

$$\int_{-\infty}^{+\infty} \mathrm{e}^{ixt} x^n \mathrm{e}^{-\frac{x^2}{2}} \frac{dx}{\sqrt{2\pi}} = n! i^n H_n(x)\mathrm{e}^{-\frac{x^2}{2}}.$$

Problem IV-32. *Let $(\Omega, \mathcal{A}, P)$ be a probability space and let $\mathcal{B}$ be a sub-σ-algebra of $\mathcal{A}$. We would like to show that if $X \in L^1(\mathcal{A})$, then*

$$(*) \qquad \int_B X dP = \int_B \mathbf{E}[X|\mathcal{B}]dP \quad \text{for all } B \in \mathcal{B},$$

and that $(*)$ characterizes $\mathbf{E}[X|\mathcal{B}]$.
(1) Show that $(*)$ holds if $X \in L^2(\mathcal{A})$.
(2) If $X > 0$, let $L(X) = \lim_{n\to+\infty} \mathbf{E}[\min(X,n)|\mathcal{B}]$. If $X \in L^1(\mathcal{A})$, let $L(X) = L(X^+) - L(X^-)$, where $X^+ = \max(X,0)$ and $X^- = \max(-X,0)$. Show that $L(X) \in L^1(\mathcal{B})$ and that $\int_B (X - L(X))dP = 0$ for all B in $\mathcal{B}$.
(3) Show that if f, $g \in L^1(\mathcal{B})$ are such that $\int_B (f-g)dP = 0$ for every B in $\mathcal{B}$, then $f = g$.
(4) Show that $L(X)$ is a bounded linear operator from $L^1(\mathcal{A})$ to $L^1(\mathcal{B})$ and infer that $L(X) = \mathbf{E}(X|B)$.

SOLUTION. (1) By definition, $X - \mathbf{E}(X|\mathcal{B})$ is orthogonal to $L^2(\mathcal{B})$ and hence, in particular, to $\mathbf{1}_B$ for every B in $\mathcal{B}$.
(2) If $X > 0$, then $\min(X,n) \in L^2(\mathcal{B})$; thus, if $B \in \mathcal{B}$,

$$\begin{aligned}\int_B X dP &= \int_B \lim_n \min(X,n)dP = \lim_n \int_B \min(X,n)dP \\ &= \lim_n \int_B \mathbf{E}[\min(X,n)|\mathcal{B}]dP = \int_B L(X)dP.\end{aligned}$$

The second equality follows from monotone convergence, the third from (1), and the fourth from the fact that $n \mapsto \mathbf{E}[\min(X,n)|\mathcal{B}]$ is an increasing sequence (see IV-2.2.1) and monotone convergence. As the limit of positive $\mathcal{B}$-measurable functions, $L(X)$ is also positive and $\mathcal{B}$-measurable. Since $\int_\Omega X dP = \int_\Omega L(X)dP$, $L(X)$ is integrable.

The case where X is not necessarily positive is clear.
(3) Without loss of generality we may assume that $g = 0$. If $B = \{f \geq \epsilon\}$, then

$$0 = \int_{B_\epsilon} f dP \geq \epsilon P(B_\epsilon);$$

thus $P(B_\epsilon) = 0$ if $\epsilon > 0$. This implies

$$P[f > 0] = P[\cup_{n=1}^{\infty} B_{\frac{1}{n}}] \leq \sum_{n=1}^{\infty} P(B_{\frac{1}{n}}) = 0,$$

and therefore $P[f > 0] = 0$. Similarly $P[-f > 0] = 0$ and $P[f = 0] = 1$.
(4) By (2) and (3), $L(X)$ is the only element of $L^1(\mathcal{B})$ such that $\int_B (X - L(X))dP = 0$ for every B in $\mathcal{B}$. It follows from (1) that $L(X) = \mathbf{E}(X|\mathcal{B})$ if $X \in L^2(\mathcal{A})$. L is linear on $L^1(\mathcal{A})$ because $\int_B (\lambda X + \mu Y - \lambda L(X) - L(Y))dP = 0$ for all B in $\mathcal{B}$. It is bounded since

$$\mathbf{E}(|L(X)|) \leq \mathbf{E}(L(X^+) + L(X^-)) = \mathbf{E}(|X|).$$

Since the extension of $\mathbf{E}(X|\mathcal{B})$ from $L^2(\mathcal{A})$ to $L^1(\mathcal{A})$ is unique (see IV-2.3), $L(X) = \mathbf{E}(X|B)$.

REMARK. This characterization of conditional expectation is often taken as a definition in the literature.

Problem IV-33. *Suppose that, for every $n \geq 0$, $X_n \in L^1(\mathcal{A})$ and $X_n \geq 0$. Use the preceding problem to show that if $X_n \uparrow X_0$, then*

$$Y_n = \mathbf{E}[X_n|\mathcal{B}] \uparrow \mathbf{E}[X_0|\mathcal{B}].$$

SOLUTION. By IV-2.2.1, $Y_n \uparrow Y$, where Y is a positive $\mathcal{B}$-measurable random variable. Hence, for every B in $\mathcal{B}$,

$$\int_B \mathbf{E}(X_0|\mathcal{B})dP = \int_B X_0 dP = \lim_{n\to\infty} \int_B X_n dP = \lim_{n\to\infty} \int_B Y_n dP = \int_B Y dP.$$

The first and third equality follow from the preceding problem, and the second and fourth from monotone convergence. Part (3) of the preceding problem implies the result.

Problem IV-34. *Suppose that $(\Omega, \mathcal{A}, P)$ is a probability space, $\mathcal{B}$ is a sub-σ-algebra of $\mathcal{A}$, Y is a $\mathcal{B}$-measurable random variable, and X is a random variable independent of $\mathcal{B}$. Consider $f : \mathbf{R}^2 \to \mathbf{R}$ such that $f(X,Y)$ is integrable. The goal of this problem is to show that if μ is the distribution of X, then*

$$(*) \qquad \mathbf{E}[f(X,Y)|\mathcal{B}] = \int_{-\infty}^{+\infty} f(x,Y)\mu(dx).$$

(1) Show that $(*)$ holds if $f(x,y) = \mathbf{1}_I(x)\mathbf{1}_J(y)$, where I and J are Borel subsets of $\mathbf{R}$.
(2) Let $\mathcal{P}$ be the Boolean algebra on $\mathbf{R}^2$ consisting of sets of the form $E = \cup_{p=1}^q I_p \times J_p$, where I_p and J_p are Borel subsets of $\mathbf{R}$. Show that $(*)$ holds if $f(x,y) = \mathbf{1}_E(x,y)$ with $E \in \mathcal{P}$.
(3) Let $\mathcal{M}$ be the family of Borel subsets M of $\mathbf{R}^2$ such that $f(x,y) = \mathbf{1}_M(x,y)$ satisfies $(*)$. Show that $\mathcal{M}$ is a monotone class.
(4) Prove $(*)$ successively for the following cases: (a) f is a simple function on $\mathbf{R}^2$; (b) f is a positive measurable function with $f(X,Y)$ integrable; and (c) the general case.

SOLUTION. (1) Follows from IV-2.3(iv), taking $\varphi = \mathbf{1}_J(Y)$.
(2) Without loss of generality we may assume the $I_p \times J_p$ disjoint; $(*)$ then follows from (1) by linearity.
(3) If $M_n \in \mathcal{M}$, with $\{M_n\}_{n=1}^\infty$ increasing and $M = \cup_{n=1}^\infty M_n$, then

$$\begin{aligned}\mathbf{E}[\mathbf{1}_M(X,Y)|\mathcal{B}] &= \lim_{n\to\infty} \mathbf{E}[\mathbf{1}_{M_n}(X,Y)|\mathcal{B}] \\ &= \lim_{n\to\infty} \int_{-\infty}^{+\infty} \mathbf{1}_{M_n}(x,Y)\mu(dx) \\ &= \int_{-\infty}^{+\infty} \mathbf{1}_M(x,Y)\mu(dx).\end{aligned}$$

The first inequality holds since (i) $\mathbf{1}_{M_n}(X,Y) \to \mathbf{1}_M(X,Y)$ in $L^2(\mathcal{A})$ by monotone convergence and (ii) the projection $L^2(\mathcal{A}) \to L^2(\mathcal{B})$ is continuous in $L^2(\mathcal{A})$. The second follows from the definition of $\mathcal{M}$ and the third from monotone convergence.

The proof of $(*)$ if $\{M_n\}_{n=1}^\infty$ is decreasing is identical to that above.
(4a) $\mathcal{M}$ is the Borel algebra on $\mathbf{R}^2$ by the theorem on monotone classes (I-1.4) applied to $\mathcal{P}$ and $\mathcal{M}$. Thus $(*)$ holds if f is a finite linear combination of indicator functions of Borel sets in $\mathbf{R}^2$.

(b) If $f \geq 0$, there exists an increasing sequence f_n of nonnegative simple Borel functions such that $f_n \uparrow f$. $(*)$ follows by an argument analogous to that of (3).

c) It suffices to write $f = f^+ - f^-$, with $f^+ = \max(0, f)$ and $f^- = \max(0, -f)$.

Problem IV-35. *On a probability space $(\Omega, \mathcal{A}, P)$, consider an integrable random variable X and a sub-σ-algebra $\mathcal{B}$ of $\mathcal{A}$, both independent of another* *sub-σ-algebra $\mathcal{C}$ of $\mathcal{A}$. Prove that if $\mathcal{D}$ is the σ-algebra generated by $\mathcal{B} \cup \mathcal{C}$, then*

$$\mathbf{E}[X|\mathcal{D}] = \mathbf{E}[X|\mathcal{B}].$$

METHOD. Prove the assertion first for square integrable X.

SOLUTION. If $X \in L^2(\mathcal{A})$, we must show that $Y = X - \mathbf{E}(X|B)$ is orthogonal not only to the subspace $L^2(\mathcal{B})$ of $L^2(\mathcal{A})$ but also to $L^2(\mathcal{D})$. For this, it suffices to show that Y is orthogonal to a dense subspace of $L^2(\mathcal{D})$. Since $\mathcal{D}$ is generated by $\{B \cap C : B \in \mathcal{B} \text{ and } C \in \mathcal{C}\}$, a dense subspace of $L^2(\mathcal{D})$ is clearly formed by the set of those $Z \in \mathcal{D}$ for which there exist (i) a $\mathcal{B}$-measurable partition $(B_1, \ldots, B_n)$ of Ω, (ii) a $\mathcal{C}$-measurable partition $(C_1, \ldots, C_m)$ of Ω, and (iii) $(a_{ij})_{i=1}^{n}{}_{j=1}^{m}$ such that $Z = \sum_{i,j} a_{ij} \mathbf{1}_{B_i} \mathbf{1}_{C_j}$. It remains to show that $\mathbf{E}(YZ) = 0$. But

$$\begin{aligned}
\mathbf{E}[YZ] &= \sum_{i,j} a_{ij} \mathbf{E}[(X - \mathbf{E}(X|B))\mathbf{1}_{B_i} \mathbf{1}_{C_j}] \\
&= \sum_{i,j} a_{ij} \mathbf{E}[(X - \mathbf{E}(X|B))\mathbf{1}_{B_i}] P[C_j],
\end{aligned}$$

since Y is independent of $\mathcal{C}$. By definition, $\mathbf{E}(Y\mathbf{1}_{B_i}) = 0$. Hence $\mathbf{E}(YZ) = 0$.

Problem IV-36. *If X and Y are integrable random variables such that $\mathbf{E}[X|Y] = Y$ and $\mathbf{E}[Y|X] = X$, show that $X = Y$ a.s.*

METHOD. Show that, for fixed x,

$$(i) \qquad 0 \leq \int_{\{Y \leq x \leq X\}} (X - Y) dP = \int_{\{x < X \text{ and } x < Y\}} (Y - X) dP,$$

and conclude by symmetry that both sides of the equation are zero. Then use Problem I-13.

SOLUTION. Since $\mathbf{E}[X|X] = X$, we can write $\mathbf{E}[X - Y|X] = 0$; thus, for every Borel set A of $\mathbf{R}$,

$$(ii) \qquad \int_{X \in A} (X - Y) dP = 0.$$

Setting $A = \{X > x\} = \{Y \leq x < X\} \cup \{x < X \text{ and } x < Y\}$, (ii) implies (i). The positivity of the left-hand side is clear. Interchanging the roles of X and Y, the same reasoning gives

$$0 \leq \int_{\{x < X \text{ and } x < Y\}} (X - Y) dP.$$

Therefore both sides of (i) are zero. Since $X-Y$ is positive on $\{Y \leq x < X\}$, we conclude that $P[Y \leq x < X] = 0$ for all real x. Problem I-13 implies that $P[X - Y > 0] = 0$. By symmetry, $P[Y - X > 0] = 0$; hence $P[X = Y] = 1$.

Problem IV-37. *Suppose that $(\Omega, \mathcal{A}, P)$ is a probability space, X and Y are integrable random variables, and $\mathcal{B}$ is a sub-σ-algebra of $\mathcal{A}$ such that X is $\mathcal{B}$-measurable.*
(1) Show that $\mathbf{E}[Y|\mathcal{B}] = X$ implies $\mathbf{E}[Y|X] = X$.
(2) Show by a counterexample that $\mathbf{E}[Y|X] = X$ does not imply that $\mathbf{E}[Y|\mathcal{B}] = X$.

SOLUTION. (1) Theorem IV-2.1.3(iii).
(2) Let X and Z be independent integrable random variables such that $\mathbf{E}(Z) = 0$ and let $\mathcal{B}$ be the σ-algebra generated by X, Z, and $Y = X + Z$. Then $\mathbf{E}[Y|X] = X + \mathbf{E}(Z) = X$ by IV-3.3.1, but

$$\mathbf{E}(Y|\mathcal{B}) = \mathbf{E}(Y|X, Z) = X + Z \neq X.$$

REMARK. If $\{\mathcal{A}_n\}_{n\geq 0}$ is a filtration of $(\Omega, \mathcal{A}, P)$, $\{X_n\}_{n\geq 0}$ a sequence adapted to this filtration, and $\mathcal{B}_n$ the σ-algebra generated by $X_0, \dots, X_n$, then $\{X_n, \mathcal{B}_n\}_{n\geq 0}$ is a martingale if $\{X_n, \mathcal{A}_n\}_{n\geq 0}$ is. The converse is false.

Problem IV-38. *Let $(Y_0, Y_1, \dots, Y_n)$ be an $(n+1)$-tuple of real random variables defined on a probability space $(\Omega, \mathcal{E}, P)$. Let $\mathcal{F}$ denote the sub-σ-algebra of $\mathcal{E}$ generated by $(Y_1^{(\omega)}, \dots, Y_n^{(\omega)}) = f(\omega)$ and assume that $\mathbf{E}(|Y_0|) < \infty$.*
(1) By applying Theorem IV-6.5.1 to f, show that there exists a Borel-measurable function $g : \mathbf{R}^n \to \mathbf{R}$ such that

$$\mathbf{E}[Y_0|\mathcal{F}] = g(Y_1, Y_2, \dots, Y_n) \quad P\text{-almost everywhere.}$$

(2) Assume that the distribution of $(Y_0, Y_1, \dots Y_n)$ in $\mathbf{R}^{n+1}$ is absolutely continuous with respect to Lebesgue measure dy_0, $dy_1, \dots, dy_n$, and let $d(y_0, y_1, \dots, y_n)$ denote its density. Prove that

$$\mathbf{E}(Y_0|\mathcal{F}) = [K(Y_1, Y_2, \dots, Y_n)]^{-1} \int_{-\infty}^{+\infty} y_0 \; d(y_0, Y_1, \dots, Y_n) dy_0,$$

where $K(y_1, \dots, y_n) = \int_{-\infty}^{+\infty} d(y_0, y_1, \dots, y_n) dy_0$. Prove that if A is a Borel subset of $\mathbf{R}$, then

$$\begin{aligned} P[Y_0 \in A|\mathcal{F}] &= \mathbf{E}[\mathbf{1}_{Y_0 \in A}|\mathcal{B}] \\ &= [K(Y_1, \dots, Y_n)]^{-1} \int_A d(y_0, Y_1, \dots, Y_n) dy_0. \end{aligned}$$

(3) Assume that the distribution of $(Y_0, Y_1, \ldots, Y_n)$ in $\mathbf{R}^{n+1}$ is Gaussian (with the definition in IV-4.3.4, which implies that $\mathbf{E}(Y_j) = 0$ for $j = 0, \ldots, n$). Use the observation that, if $(X, Y_1, \ldots, Y_n)$ is Gaussian in $\mathbf{R}^{n+1}$, then X is independent of $(Y_1, \ldots, Y_n)$ if and only if $\mathbf{E}(XY_j) = 0$ $\forall j = 1, \ldots, n$, to show that there exist real numbers $\lambda_1, \ldots, \lambda_n$ such that $\mathbf{E}[Y_0|\mathcal{F}] = \lambda_1 Y_1 + \cdots + \lambda_n Y_n$.

SOLUTION. (1) Let $\mathcal{A}'$ be the sub-σ-algebra of $\mathcal{E}$ generated by $Y_0, Y_1, \ldots, Y_n$ and let $\mathcal{A}$ be the completion of $\mathcal{A}'$ with respect to P. Then $(\Omega, \mathcal{A}, P)$ is complete and separable. We will apply Theorem IV-6.5.1 to $(\Omega, \mathcal{A}, P)$, $h(\omega) = Y_0(\omega)$, and $(X, \mathcal{B}) = \mathbf{R}^n$ equipped with its Borel algebra. Define

$$g(y_1, \ldots, y_n) = \int_\Omega h(\omega) d\nu_{y_1,\ldots,y_n}(\omega).$$

If $v \in L^\infty(\mathcal{F})$, we know by Dynkin's theorem (IV-1.5.4) that there exists $u \in L^\infty(\mathcal{B})$ such that if μ is the distribution of $(Y_1, \ldots, Y_n)$, then

$$\mathbf{E}[vY_0] = \int_{\mathbf{R}} u(y_1, \ldots, y_n) g(y_1, \ldots, y_n) \mu(dy_1, \ldots, dy_n).$$

But this is equivalent to the statement $\mathbf{E}[Y_0|E] = g[Y_1, \ldots, Y_n]$.
(2) Let $u \in L^\infty(\mathcal{B})$ and let $v = u \circ f$. Then

$$\begin{aligned}&\int_{\mathbf{R}^n} u(y_1, \ldots, y_n) \left(\int_{-\infty}^{+\infty} y_0 d(y_0, y_1, \ldots, y_n) dy_0 \right) \cdot \\ &\qquad (K(y_1, \ldots, y_n))^{-1} \mu(dy_1, \ldots, dy_n) \\ &= \int_{\mathbf{R}^{n+1}} y_0 u(y_1, \ldots, y_n) d(y_0, y_1, \ldots, y_n) dy_0 dy_1 \ldots dy_n = \mathbf{E}[Y_0 v],\end{aligned}$$

since $\mu(dy_1, \ldots, dy_n) = K(y_1, \ldots, y_n) dy_1 \ldots dy_n$. This completes the proof. The case $P[Y_0 \in A|\mathcal{F}]$ is similar.
(3) Consider the subspace G of $L^2(\mathcal{F})$ (real) consisting of random variables of the form $\alpha_1 Y_1 + \cdots + \alpha_n Y_n$, with $\alpha_1, \ldots, \alpha_n \in \mathbf{R}$. G is closed since $\dim G \leq n < \infty$; let $\lambda_1 Y_1 + \cdots \lambda_n Y_n$ denote the orthogonal projection of Y_0 on G. (The $\lambda_1, \ldots, \lambda_n$ are not necessarily unique; they are unique if and only if $\det(\mathbf{E}(Y_i Y_j))_{i,j=1}^n \neq 0$.) If $X = Y_0 - (\lambda_1 Y_1 + \cdots \lambda_n Y_n)$, then the distribution of $(X, Y_1, \ldots, Y_n)$ is a Gaussian distribution on $\mathbf{R}^{n+1}$. Since $\mathbf{E}[XY_j] = 0$ for $j = 1, \ldots, n$ by the construction of X, X is independent of $(Y_1, \ldots, Y_n)$; that is, X is independent of the σ-algebra $\mathcal{F}$.Hence $\mathbf{E}(X|\mathcal{F}) = \mathbf{E}[X] = 0$, and it follows that $\mathbf{E}[Y_0|\mathcal{F}] = \mathbf{E}[\lambda_1 Y_1 + \cdots \lambda_n Y_n|\mathcal{F}] = \lambda_1 Y_1 + \cdots \lambda_n Y_n$, since Y_j is $\mathcal{F}$-measurable.

Problem IV-39. *Let $\{X_n\}$ be a sequence of independent real random variables with the same distribution and let $\mathcal{F}_n$ be the σ-algebra generated by $X_1, \ldots, X_n$. Set $S_n = X_1 + \cdots + X_n$ for $n > 0$ and set $S_0 = 0$. Which of the following processes are martingales relative to the filtration $\{\mathcal{F}_n\}_{n=0}^\infty$?*

(1) S_n, *if* $\mathbf{E}(|X_1|) < \infty$.
(2) $X_1^2 + \cdots + X_n^2 - n\lambda$, *if* $\mathbf{E}(X_1^2) < \infty$ *and* λ *is real.*
(3) $\exp(\alpha S_n - n\lambda)$, *if* $\varphi(\alpha) = \mathbf{E}(\exp(\alpha X_1)) < \infty$ *and* α *and* λ *are real.*
(4) $Y_n = |S_{\min(n,T)}|$, *where* $T = \inf\{n > 0 : S_n = 0\}$, *and we assume that* $P[X_1 = 1] = P[X_1 = -1] = \frac{1}{2}$.

SOLUTION. (1) $\mathbf{E}(S_n|\mathcal{F}_n) = S_n$ since S_n is $\mathcal{F}$-measurable. $\mathbf{E}(X_{n+1}|\mathcal{F}_n) = \mathbf{E}(X_{n+1})$ since X_{n+1} is independent of $\mathcal{F}_n$. Thus $\mathbf{E}(S_{n+1}|\mathcal{F}_n) = S_n + \mathbf{E}(X_1)$, and $\{S_n\}_{n=0}^{\infty}$ is a martingale if and only if $\mathbf{E}(X_1) = 0$.
(2) Similarly,

$$\mathbf{E}[X_1^2 + \cdots + X_{n+1}^2 - (n+1)\lambda|\mathcal{F}_n] = X_1^2 + \cdots + X_n^2 - n\lambda + (\mathbf{E}(X_1^2) - \lambda),$$

and we have a martingale if and only if $\lambda = \mathbf{E}(X_1^2)$.
(3) $\mathbf{E}[\exp\alpha(S_n + X_{n+1})|\mathcal{F}_n] = \exp(\alpha S_n)\mathbf{E}[\exp(\alpha X_{n+1})|\mathcal{F}_n] = \varphi(\alpha)\exp(\alpha S_n)$. We have a martingale if and only if $\lambda = \log\varphi(\alpha)$.
(4) In this case, the σ-algebra is generated by the finite partition consisting of $A(\epsilon) = \{X_1 = \epsilon_1, X_2 = \epsilon_2, \ldots, X_n = \epsilon_n\}$, where the $\epsilon_j = \pm 1$. If $\omega \in A(\epsilon)$, then

$$\mathbf{E}(Y_{n+1}|\mathcal{F}_n)(\omega) = \mathbf{E}(Y_{n+1}\mathbf{1}_{A(\epsilon)})\frac{1}{P(A(\epsilon))}.$$

If $T(\omega) \leq n$, then $Y_n(\omega) = 0$, and $Y_{n+1} = 0$ on $A(\epsilon)$. If $T(\omega) > n$, then $S_n(\omega) \neq 0$ and

$$\begin{aligned} Y_{n+1} &= Y_n + X_{n+1} \quad \text{if} \quad S_n > 0 \\ Y_{n+1} &= Y_n - X_{n+1} \quad \text{if} \quad S_n < 0, \end{aligned}$$

since S_n is integer valued. Hence

$$\mathbf{E}[Y_{n+1}|\mathcal{F}_n](\omega) = Y_n(\omega) \quad \text{if } T(\omega) > n.$$

Therefore Y_n is a martingale.

Problem IV-40. *Let* $Y_1, \ldots, Y_n, \ldots$ *be independent real random variables with the same distribution and such that* $\mathbf{E}[|Y_1|] < \infty$. *Set* $S_n = Y_1 + \cdots + Y_n$.
(1) Show that $\mathbf{E}[Y_k|S_n] = \frac{S_n}{n}$ *if* $1 \leq k \leq n$.
(2) If m *is fixed and* $X_k = \frac{S_{m-k}}{m-k}$ *for* $0 \leq k \leq m-1$, *show that* $(X_0, \ldots, X_{m-1})$ *is a martingale. (Apply Problem IV-35.)*

SOLUTION. It is clear by symmetry that $\mathbf{E}[Y_1|S_n] = \mathbf{E}[Y_2|S_n] = \ldots = \mathbf{E}[Y_n|S_n]$. For a rigorous proof of this fact, consider the probability space $\mathbf{R}^n$ equipped with the measure $\otimes_1^n \nu_i$, where ν_i is the distribution of Y_i. Next, consider the action of the group Σ_n of permutations of $\{1, \ldots, n\}$ on $\mathbf{R}^n$ defined by $\sigma(y_1, \ldots, y_n) = (y_{\sigma(1)}, \ldots, y_{\sigma(n)})$ if $\sigma \in \Sigma_n$. It is clear that σ preserves both $\otimes_1^n \nu_i$ and $S_n = y_1 + \cdots y_n$. Hence

$$\mathbf{E}[Y_k|S_n] = \frac{1}{n}\sum_{j=1}^{n}\mathbf{E}[Y_j|S_n] = \frac{1}{n}\mathbf{E}[\sum_{j=1}^{n} Y_j|S_n] = \frac{1}{n}S_n.$$

(2)

$$\begin{aligned}&\mathbf{E}[X_k|X_0, X_1, \dots, X_{k-1}] = \\ &= \tfrac{1}{m-k}\mathbf{E}[Y_1 + \cdots Y_{m-k}|Y_1 + \cdots Y_{m-k+1}, Y_{m-k+2}, \dots, Y_m].\end{aligned}$$

Indeed, the σ-algebra $\mathcal{D}_{k-1}$ generated by $X_0, \dots, X_{k-1}$ is the same as the σ-algebra generated by $X_{k-1}, Y_{m-k+2}, \dots, Y_{m-1}, Y_m$. Let $\mathcal{B}_{k-1}$ be the σ-algebra generated by X_{k-1} and let $\mathcal{C}_{k-1}$ be the σ-algebra generated by $Y_{m-k+2}, \dots, Y_{m-1}, Y_m$. Then X_k and $\mathcal{B}_{k-1}$ are independent of the σ-algebra $\mathcal{C}_{k-1}$, and

$$\begin{aligned}\mathbf{E}[X_k|\mathcal{D}_{k-1}] &= \mathbf{E}[X_k|\mathcal{B}_{k-1}] && \text{by Problem IV-35} \\ &= X_{k-1} && \text{by (1).}\end{aligned}$$

Problem IV-41. *Let $\{X_n\}_{n=1}^{\infty}$ be a sequence of independent random variables with the same distribution defined by $P[X_n = k] = 2^{-k}$ for $k = 1, 2, \dots$. Random variables Z_n are defined by letting Z_0 be a positive constant and setting $Z_n = \frac{3Z_{n-1}}{2^{X_n}}$ for $n = 1, 2, \dots$.*
(1) Prove that $\{Z_n\}_{n=0}^{\infty}$ is a martingale relative to the filtration $\{\mathcal{F}_n\}_{n=0}^{\infty}$, where $\mathcal{F}_n$ is the σ-algebra generated by $X_1, \dots, X_n$.
(2) Use the law of large numbers (see Problem IV-6) to prove that $Z_n \to 0$ almost surely as $n \to \infty$.

SOLUTION. (1) Observe that $\mathbf{E}\left(\frac{1}{2^{X_n}}\right) = \frac{1}{4} + \frac{1}{4^2} + \cdots = \frac{1}{3}$. It follows that

$$\mathbf{E}[Z_n|\mathcal{F}_{n-1}] = 3Z_{n-1}\mathbf{E}\left(\frac{1}{2^{X_n}}\right) = Z_{n-1}.$$

(2) $Z_n = Z_0 2^{-(X_1+X_2+\cdots+X_n)}$. Hence, by the strong law of large numbers,

$$\frac{1}{n}\log Z_n = -\frac{X_1 + \cdots + X_n}{n}\log 2 \to -2\log 2 \text{ a.s. as } n \to \infty.$$

This implies that $Z_n \to 0$ a.s. as $n \to \infty$.

REMARK. This gives a heuristic confirmation of the following unproved conjecture in number theory. If n is an odd positive integer, let $f(n) = (3n+1)2^{-\nu(3n+1)}$, where $2^{\nu(3n+1)}$ denotes the greatest power of 2 that divides the integer $3n+1$. The conjecture asserts that, for every n, there exists an integer k such that the kth iterate of f satisfies $f^{(k)}(n) = 1$. If n is very large, $\nu(3n+1)$ appears to behave like the variable X_1 of the problem, and $\{Z_k\}_{k=1}^{\infty}$ like the sequence $\{f_k(n)\}_{k=1}^{\infty}$.

Problem IV-42. *Let $H \subset L^1(\Omega, \mathcal{A}, P)$, where $(\Omega, \mathcal{A}, P)$ is a probability space.*

(1) If F is a positive function on $(0,+\infty)$ such that $\frac{F(x)}{x}$ is increasing and $\to +\infty$ as $n \to \infty$, and if

$$\sup_{h\in H} \mathbf{E}(F|h) = M < \infty,$$

show that H is uniformly integrable.
METHOD. Use Proposition IV-5.7.2.

(2) If H is a bounded subset of $L^p(\Omega, \mathcal{A}, P)$ with $p > 1$, show that H is uniformly integrable.

SOLUTION. (1) If $q > 0$ and $h \in H$, then

$$\int_{|h|>q} |h| dP \le \frac{q}{F(q)} \int_{|h|>q} F(|h|) dP \le \frac{qM}{F(q)} \to 0 \quad \text{as } q \to \infty.$$

(2) Set $M = \sup \|h\|^p_{L^p}$ and $F(x) = x^p$. Then the result follows from (1).

Problem IV-43. *Let $\{X_{ij}\}_{i,j=1}^{\infty}$ be independent random variables with values in $\mathbf{N}$ and with the same distribution. Assume that $0 < m = \mathbf{E}(X_{11}) < \infty$ and that $\sigma^2 = \mathbf{E}((X_{11} - m)^2) < \infty$. Consider the sequence of random variables defined by*

$$\begin{array}{lll} Z_0 = 1 & & \\ Z_{n+1} = 0 & \text{if} & Z_n = 0 \\ Z_{n+1} = \sum_{i=1}^{Z_n} X_{i,n+1} & \text{if} & Z_n > 0. \end{array}$$

$\mathcal{F}_n$ is the σ-algebra generated by $\{X_{i,j} : 1 \le i < \infty,\ 1 \le j \le n\}$.
(1) Show that $\{\frac{Z_n}{m^n}, \mathcal{F}_n\}_{n=1}^{\infty}$ is a martingale.
(2) Show that $\mathbf{E}\left(\frac{Z_{n+1}^2}{m^{2(n+1)}}\right) = \mathbf{E}\left(\frac{Z_n^2}{m^{2n}}\right) + \frac{\sigma^2}{m^{2n+1}}$.

Conclude that, if $m > 1$, the martingale is regular. (Use Problem IV-42 and Theorem IV-5.8.1.)

SOLUTION. (1) It is clear that $\mathbf{E}[Z_{n+1}|\mathcal{F}_n] = \mathbf{E}[Z_{n+1}|Z_n] = mZ_n$, which implies the result.

(2)

$$\begin{aligned} \mathbf{E}[Z_{n+1}^2] &= \mathbf{E}[\mathbf{E}(Z_{n+1}^2|Z_n)] \\ &= \mathbf{E}[(\mathbf{E}(Z_{n+1}|Z_n))^2 + \mathbf{E}([Z_{n+1} - \mathbf{E}(Z_{n+1}|Z_n)]^2|Z_n] \\ &= \mathbf{E}(m^2 Z_n^2 + \sigma^2 Z_n) \\ &= m^2 \mathbf{E}(Z_n^2) + \sigma^2 m, \end{aligned}$$

as claimed. Hence, if $m > 1$,

$$\begin{aligned} \mathbf{E}\left(\frac{Z_n^2}{m^{2n}}\right) &= \mathbf{E}\left(\frac{Z_0^2}{m^0}\right) + \sum_{k=0}^{n-1} \left[\mathbf{E}\left(\frac{Z_{k+1}^2}{m^{2(k+1)}}\right) - \mathbf{E}\left(\frac{Z_k^2}{m^{2k}}\right)\right] \\ &< 1 + \frac{\sigma^2}{m}\left(1 - \frac{1}{m^2}\right)^{-1}. \end{aligned}$$

The martingale is uniformly integrable by Problem IV-42, and regular by Theorem IV-5.8.1.

REMARK. $\{Z_n\}_{n=0}^{\infty}$ is sometimes called the Galton-Watson process, and serves as a model in genetics. ($X_{i,j}$ is the number of offspring of the individual i of the jth generation, which has total size Z_j.)

V

Gaussian Sobolev Spaces and Stochastic Calculus of Variations

Problem V-1. *Let E be the set of compactly supported C^∞ functions on* $\mathbf{R}$, *and let d and δ be the operators on E defined by*

$$(d\varphi)(x) = \varphi'(x) \quad \textit{and} \quad (\delta\varphi)(x) = -\varphi'(x) + x\varphi(x).$$

(1) Prove by induction on n that

$$d^n\delta - \delta d^n = nd^{n-1}.$$

(2) Let p be a norm on E. Let B be the algebra of operators on E which are continuous with respect to this norm; that is, the set of endomorphisms a of E such that

$$\|a\| = \sup\{p(a(\varphi)) : p(\varphi) \leq 1\}$$

is finite. Assume that d and δ are in B. Use (1) to prove that, for all $n \geq 1$,

$$\|nd^{n-1}\| \leq 2\|d^{n-1}\| \; \|d\| \; \|\delta\|.$$

(3) Deduce from (2) that d and δ are never simultaneously continuous.

SOLUTION. (1) The result holds for $n = 1$ by V-1.3.2(iii). If it holds for n, then multiplying the identity on the left by d gives

$$nd^n = d^{n+1}\delta - (d\delta)d^n = d^{n+1}\delta - d^n - \delta d^{n+1},$$

where the last equality follows from the equation $d\delta - \delta d =$ identity.
(2) It is standard to check that $a \mapsto \|a\|$ is a norm on B and that $\|ab\| \leq \|a\| \; \|b\|$. Hence

$$\|nd^{n-1}\| \leq \|d^n\delta\| + \|\delta d^n\| \leq 2\|d^{n-1}\| \; \|d\| \; \|\delta\|.$$

(3) $d^{n-1} \neq 0$ for all $n \geq 1$ by the definition of d. Hence, by (2), $n \leq 2\|d\| \, \|\delta\|$ for all n — a contradiction.

REMARK. This result is due to Aurel Wintner (1947).

Problem V-2. *Let $\{H_n\}_{n=0}^{\infty}$ be the sequence of Hermite polynomials defined in V-1.3.*
(1) Use Proposition V-1.3.4 to show that, for $n \geq 1$,

$$H_{n+1} + nH_{n-1} = xH_n.$$

If $\widetilde{H}_n = \frac{H_n}{n!}$ (compare with Problem IV-29), show that

$$\widetilde{H}_{n+1} = \frac{x}{n+1}\widetilde{H}_{n+1} - \frac{1}{n+1}\widetilde{H}_{n-1}.$$

(2) Conclude from (1) that the radius of convergence $R(x)$ of $\sum_{n=0}^{\infty} \frac{z^n}{n!} H_n(x)$ is $+\infty$ for every complex number x.

METHOD. For (2), show that for every $\epsilon > 0$ there exist an integer $N(\epsilon)$ and a sequence $\{x_n\}_{n=N(\epsilon)-1}^{\infty}$ such that $|\widetilde{H}_n| \leq x_n$ and $x_{n+1} = \epsilon x_n + \epsilon x_{n-1}$.

SOLUTION. (1) By the proposition, $H_{n+1} = \delta H_n \;\; nH_{n-1} = dH_n$, and $xH_n = (\delta + d)H_n$. This proves the first statement; writing $H_k = k!\widetilde{H}_k$ proves the second.
(2) By (1), for every $\epsilon > 0$ there exists $N(\epsilon)$ such that $n \geq N(\epsilon)$ implies

$$|\widetilde{H}_{n+1}| \leq \epsilon|\widetilde{H}_n| + \epsilon|\widetilde{H}_{n-1}|.$$

Define a sequence $\{x_n\}$ by $x_n = |\widetilde{H}_n|$ for $n = N(\epsilon) - 1$ and $n = N(\epsilon)$, and $x_{n+1} = \epsilon x_n + \epsilon x_{n-1}$ for $n > N(\epsilon)$. Then it is clear, by induction on n, that $|\widetilde{H}_n| \leq x_n$ for all $n \geq N(\epsilon)$.

By the elementary theory of sequences satisfying a linear recursion relation, if r_+ and r_- are the two roots of the equation $r^2 - \epsilon r + \epsilon$, there exist two numbers A_+ and A_- such that

$$x_n = A_+ r_+^n + A_- r_-^n.$$

Therefore the radius of convergence R of the series $\sum_n z^n x_n$ is such that $Rr_+ \geq 1$ and $R|r_-| \geq 1$. Since $R \leq R(x)$ and r_+ and r_- both approach 0 as $\epsilon \to 0$, it is clear that $R(x) = +\infty$.

Problem V-3. *Let $\{H_n\}_{n=0}^{\infty}$, d, and δ be defined as in V-1.3. For nonnegative integers n, consider*

$$F_n(x) = H_n(ix)(-i)^n.$$

Let $\lambda \in \mathbf{C}$ and define ρ by $\rho = \delta + \lambda d$.

(1) For $n \geq 1$, prove that $d^n \rho = \rho d^n + n d^{n-1}$ and $F_{n+1} = xF_n + nF_{n-1}$.
(2) Prove by induction on n that

$$(d+\rho)^n = \sum_{k=0}^{n} C_n^k H_k(\rho) d^{n-k},$$

where C_n^k denotes the binomial coefficient.
(3) If φ is a polynomial and t is real, let $\tau_t(\varphi)(x) = \varphi(x+t)$. Prove that $(\exp(td))(\varphi) = \tau_t(\varphi)$ *and that*

$$(\exp\ t(d+\rho))(\varphi) = \left(\exp\frac{t^2}{2}\exp(t\rho)\tau_t\right)(\varphi).$$

In particular, if $\lambda = 1$ (that is, if $\rho(\varphi)(x) = x\varphi(x)$), compute $(\exp\ t(d+x))(\varphi)$.

SOLUTION. (1) Since $d\delta - \delta d =$ identity, the first identity to be proved is clear for $n=1$. For arbitrary n, it can be proved by induction as in Problem V-1(1). The second identity is proved by writing $H_n(ix) = i^n F_n(x)$ and using the fact that $H_{n+1}(x) + nH_{n-1}(x) = xH_n(x)$ (Proposition V-1.3.4).
(2) Equality is obvious for $n=0$. Assume that it holds for n and multiply on the right by $d+\rho$. Setting

$$A = \sum_{k=0}^{n} C_n^k F_k(\rho) d^{n+1-k} \quad \text{and} \quad B = \sum_{k=0}^{n} C_n^k F_k(\rho) d^{n-k}\rho$$

gives $(d+\rho)^n = A + B$. Next, re-write B using (1) applied to $n-k$:

$$B = \sum_{k=0}^{n} C_n^k F_k(\rho)(\rho d^{n-k} + (n-k)d^{n-k-1}).$$

The change of indices $k' = k+1$ and $k'' = k+2$ gives

$$B = \sum_{k=1}^{n+1} C_n^{k-1}\rho F_{k-1}(\rho) d^{n+1-k} + \sum_{k=2}^{n+1} C_n^{k-2}(n-k+2)F_{k-2}(\rho)d^{n+1-k}.$$

Thus $A+B$ is equal to

$$\sum_{k=2}^{n+1}[C_n^k F_k(\rho) + C_n^{k-1}\rho F_{k-1}(\rho) + C_n^{k-2}(n-k+2)F_{k-2}(\rho)]d^{n+1-k} + \\ d^{n+1} + nF_1(\rho)d^n + \rho d^n + [\rho F_n(\rho) + nF_{n-1}(\rho)].$$

For $k = 2, 3, \ldots, (n+1)$, use (1) and replace $\rho F_{k-1}(\rho)$ by $F_k(\rho) - (k-1)F_{k-2}$. Recalling that $F_1(\rho) = \rho$, we find that

$$A + B = \sum_{k=2}^{n}(C_n^k + C_n^{k-1})F_k(\rho)d^{n+1-k} + d^{n+1} + (n+1)\rho d^n + F_{n+1}(\rho).$$

This is the desired formula, since $C_n^k + C_n^{k-1} = C_{n+1}^k$.
(3) $(\exp(td))(\varphi)(x) = \sum_{n=0}^{\infty} \frac{t^n}{n!}\varphi^{(n)}(x) = \varphi(t+x)$ by Taylor's formula. It follows from (2) that

$$\begin{aligned}(\exp\ t(d+\rho))(\varphi)(x) &= \sum_{n=0}^{\infty}\frac{t^n}{n!}(d+\rho)^n\varphi(x)\\ &= \sum_{n=0}^{\infty}\sum_{k=0}^{n}\frac{t^k}{k!}F_k(\rho)\frac{t^{n-k}}{(n-k)!}\varphi^{(n-k)}(x).\end{aligned}$$

But, by the definition of the F_k and formula V-1.4.2,

$$\sum_{k=0}^{\infty}\frac{t^k}{k!}F_k(x) = \exp\left[\frac{t^2}{2} + tx\right].$$

Thus, using the formula for the Cauchy product of two power series,

$$\begin{aligned}(\exp\ t(d+\rho))(\varphi)(x) &= \left(\sum_{n=0}^{\infty}\frac{t^k}{k!}F_k(\rho)\right)\left(\sum_{n=0}^{\infty}\frac{t^n}{n!}\varphi^{(n)}(x)\right)\\ &= \left(\exp(\tfrac{t^2}{2} + t\rho)\tau_t\right)(\varphi)(x).\end{aligned}$$

In particular, if $\lambda = 1$ then $\rho(\varphi) = x\varphi(x)$ and

$$(\exp t(d+x))(\varphi)(x) = \exp(\frac{t^2}{2} + tx)\varphi(t+x).$$

REMARK. The result of (2) is due to Viskov[1]; that of (3) is due to Ville[2].

Problem V-4. *Let X and Y be independent random variables with the same distribution $\nu_1(dx) = \exp(-\frac{x^2}{2})\frac{dx}{\sqrt{2\pi}}$. Let $g : \mathbf{R} \to [0, +\infty)$ be a measurable function and let $Z = X + Y\sqrt{g(X)}$. Assume that Z has a normal distribution. Cantelli conjectured in 1917 that g is then constant almost everywhere; this is still unproved in 1994.*
(1) Let $g_0 = \mathbf{E}(g(X))$. For all real t, compute $\mathbf{E}(\exp\ tZ)$ as a function of g_0. Prove that $\exp(\alpha g) \in L^2(\nu_1)$ for all $\alpha > 0$.
METHOD. Use the Schwarz inequality.

(2) Let $\{g_n\}_{n=0}^{\infty}$ be the sequence of real numbers such that $g(x) = \sum_{n=0}^{\infty} g_n\frac{H_n(x)}{n!}$ in the $L^2(\nu_1)$ sense. By considering $\mathbf{E}(Z^3)$ and $\mathbf{E}(Z^4)$, show that $g_1 = 0$ and $-2g_2 = \sum_{n=2}^{\infty}\frac{g_n^2}{n!}$.

---

[1]O. Viskov, *Theory of Probability and Its Applications*, Vol. 30, n. 1 (1984), 141–143.

[2]J. Ville, *Comptes Rendus Acad. des Sc. 221* (1945), 529–539.

(3) Prove that $g(x) \leq g_0 + 1$ almost everywhere.
METHOD. If $\epsilon > 0$, let $A_\epsilon = \{x : g(x) \geq \epsilon + g_0 + 1\}$ and let a be a real number such that $A'_\epsilon = A_\epsilon \cap [a, +\infty)$ has positive measure. Consider

$$\int_{A'_\epsilon} \exp[tx + \frac{t^2}{2}(g(x) - 1 - g_0)]\nu_1(dx).$$

SOLUTION. (1) $\mathbf{E}(Z^2) = \mathbf{E}(X^2 + 2YX\sqrt{g(X)} + Y^2 g(X))$. Hence $\mathbf{E}(\exp\ tZ) = \exp[\frac{t^2}{2}(1+g_0)]$. But

$$\mathbf{E}(\exp\ tZ) = \mathbf{E}[\exp(tX + tY\sqrt{g(X)}] = \mathbf{E}[\exp(tX + t^2\frac{g(X)}{2})].$$

Applying Schwarz's inequality to the product of $\exp[\frac{1}{2}(tX + t^2\frac{g(X)}{2})]$ and $\exp(-\frac{tX}{2})$ gives

$$\left[\mathbf{E}\left(\exp\frac{t^2 g(X)}{4}\right)\right]^2 \leq \exp(\frac{t^2}{2}(1+g_0))\exp(\frac{t^2}{2}).$$

Hence, for $\alpha \geq 0$,

$$\mathbf{E}[\exp(\alpha g(X))] \leq \exp(\alpha(2 + g_0)).$$

(2) Define a sequence $\{g_n\}$ by $\frac{g_n}{\sqrt{n!}} = \langle g, \frac{H_n}{\sqrt{n!}}\rangle$; that is, $g_n = \mathbf{E}(g(X)H_n(X))$. Then, since $\{\frac{H_n}{\sqrt{n!}} : n \geq 0\}$ is a basis for the Hilbert space $L^2(\nu_1)$,

$$\sum_{n=0}^{\infty} \frac{g_n^2}{n!} < +\infty.$$

Next,

$$\mathbf{E}(Z^3) = \mathbf{E}[X^3 + 3YX^2\sqrt{g(X)} + 3XY^2 g(X) + Y^3(\sqrt{g(X)})^3].$$

Since the odd moments of the centered Gaussian variables X, Y, and Z are zero and $H_1(X) = X$, we see that $g_1 = \mathbf{E}(Xg(X)) = 0$. Similarly,

$$\mathbf{E}(Z^4) = \mathbf{E}(X^4 + 6X^2Y^2 g(X) + Y^4 g^2(X)).$$

Since $\mathbf{E}(X^4) = 3$ and therefore $\mathbf{E}(Z^4) = 3(1+g_0)^2$, we find that $(1+g_0)^2 = 1 + 2\mathbf{E}[X^2 g(X)] + \mathbf{E}[(g(X))^2]$. The desired result follows since $X^2 = 1 + H_2(X)$ and

$$\mathbf{E}[(g(X))^2] = g_0^2 + \sum_{n=2}^{\infty} \frac{g_n^2}{n!}.$$

(3)

$$\begin{array}{rcl} 1 & = & \mathbf{E}[\exp(tZ - \frac{t^2}{2}(1+g_0))] \\ & = & \mathbf{E}[\exp(tX + \frac{t^2}{2}(g(X) - 1 - g_0))] \\ & \geq & \mathbf{E}[\mathbf{1}_{A'_\epsilon} \exp(tX + \frac{t^2}{2}(g(X) - 1 - g_0))] \\ & \geq & \exp(ta + \frac{t^2}{2}\epsilon)P(X \in A'_\epsilon). \end{array}$$

As $t \to +\infty$, this gives a contradiction if $P(X \in A'_\epsilon) > 0$.

Problem V-5. *As usual, we denote by $\{H_n\}_{n\geq 0}$ the sequence of Hermite polynomials and by ν_1 the normal distribution on $\mathbf{R}$. Let μ be a probability distribution on $\mathbf{R}^2$ such that if (X,Y) has distribution μ, then X and Y have distribution ν_1 and there exists a real sequence $\{C_n\}_{n\geq 0}$ with*

$$\mathbf{E}(H_n(X)|Y) = C_n H_n(Y).$$

(1) Prove that $C_n = \mathbf{E}(H_n(X)H_n(Y))$ and $-1 \leq C_n \leq 1$ for all n in $\mathbf{N}$.
(2) Prove that if $\sum_{n\geq 1} C_n^2 < +\infty$, then μ is absolutely continuous with respect to $\nu_1(dx)\nu_1(dy)$ and its density is

$$f(x,y) = \sum_{n\geq 0} \frac{C_n}{n!} H_n(x)H_n(y).$$

METHOD. For (2), write $\mu(dx,dy) = \nu_1(dy)K(y,dx)$. Show that the function $x \mapsto f(x,y)$ is in $L^2(\nu_1)$ y-almost everywhere and that, for every $\theta \in \mathbf{C}$,

$$\int \exp(\theta x)(f(x,y)\nu_1(dx) - K(y,dx)) = 0 \quad y\text{-a.e.}$$

SOLUTION.(1)

$$\begin{array}{rcl} \mathbf{E}(H_n(X)H_n(Y)) & = & \mathbf{E}[\mathbf{E}(H_n(X)H_n(Y)|Y)] \\ & = & \mathbf{E}[H_n(Y)\mathbf{E}(H_n(X)|Y)] \\ & = & C_n\mathbf{E}[(H_n(Y))^2] = n!C_n, \end{array}$$

by the rules for computing conditional expectations. The Schwarz inequality then shows that $|C_n| \leq 1$.
(2) $\sum_{n=0}^{\infty} C_n^2 < +\infty$ implies that f is in $L^2(\nu_1 \otimes \nu_1)$, since

$$\int\int (f(x,y))^2 \nu_1(dx)\nu_1(dy) = \sum_{n=0}^{\infty} C_n^2.$$

Thus, by Fubini's theorem,

$$\int (f(x,y))^2 \nu_1(dx) < +\infty$$

$\nu_1(dy)$-almost surely and hence y-almost everywhere. Similarly, if θ is real,

$$\int\int e^{\theta x}|f(x,y)|\nu_1(dx)\nu_1(dy) < +\infty$$

since $(x,y) \mapsto e^{\theta x}$ is in $L^2(\nu_1 \otimes \nu_1)$ and therefore $\int e^{\theta x}|f(x,y)|\nu_1(dx) < +\infty$ y-almost everywhere. Moreover, since X has distribution ν_1,

$$\exp\left(\frac{\theta^2}{2}\right) = \int\int e^{\theta x}\mu(dx,dy) = \int \nu_1(dy) \int e^{\theta x}K(y,dx),$$

and $\int e^{\theta x}K(y,dx) < +\infty$ y-almost everywhere. Thus

$$\int H_n(x)f(x,y)\nu_1(dx) = C_n H_n(y) = \int H_n(x)K(y,dx)$$

and, since the $\{H_n : n \geq 0\}$ form a basis for the polynomials,

$$\int x^n(f(x,y)\nu_1(dx) - K(y,dx)) = 0 \quad \forall n \in \mathbf{N}.$$

Finally, since $\int e^{|\theta x|}(|f(x,y)|\nu_1(dy) + K(y,dx)) < +\infty$, we conclude that

$$\int \exp(\theta x)f(x,y)\nu_1(dx) = \int \exp(\theta x)K(y,dx)$$

for all complex θ. This implies that the bounded signed measure $f(x,y)\nu_1(dx)$ coincides with the probability measure $K(y,dx)$ y-almost everywhere.

Problem V-6. *We keep the notation of Problem V-5 and denote by C the set of probability measures μ on $\mathbf{R}^2$ described there. Let μ be a fixed element of C.*
(1) Define $\{b_{n,k}\}_{0 \leq k \leq n}$ by

$$x^n = \sum_{k=0}^{n} b_{n,k}H_k(x)$$

and let

$$P_n(y) = \sum_{k=0}^{n} b_{n,k}C_k H_k(y).$$

Show that $\int x^n K(y,dx) = P_n(y)$ y-a.e. and that $\lim_{y\to\infty} y^{-n}P_n(y) = C_n$.
(2) Let $\sigma(y,dt)$ be the image of $K(y,dx)$ under the mapping $x \mapsto \frac{x}{y}$. For $\theta \in \mathbf{C}$, show that

$$\int_{-\infty}^{+\infty} \exp(\theta t)\sigma(y,dt) = \exp\left(\frac{\theta^2}{2y^2}\right)\sum_{k=0}^{\infty}\frac{C_k}{k!}\theta^k y^{-k}H_k(y)$$

and

$$\lim_{y\to\infty}\int_{-\infty}^{+\infty}\exp(\theta t)\sigma(y,dt)=\sum_{k=0}^{\infty}\frac{C_k}{k!}\theta^k.$$

(3) Show that the probability measure $\sigma(dt)=\lim_{y\to\infty}\sigma(y,dt)$ exists and that

$$C_n=\int t^n\sigma(dt).$$

From the fact that $|C_n|\leq 1$, conclude that $\sigma(\mathbf{R}\setminus[-1,1])=0$.
(4) Show that σ is the unique probability measure on $[-1,1]$ such that $C_n=\int_{-1}^{1}t^n\sigma(dt)$.
(5) Show that the mapping $\mu\mapsto\sigma$, from $\mathcal{C}$ to the set of probability measures on $[-1,1]$, is a bijection. What is μ when σ is the Dirac measure at ρ?

METHOD. For (5), consider successively the cases where $\rho=1$, $\rho=-1$, and (using Problem V-5(2) and Mehler's formula, V-1.5.8(ii)) $\rho<1$.
SOLUTION. (1) $\int H_k(x)K(y,dx)=C_kH_k(y)$ y-almost everywhere; the definition of P_n implies the first assertion. The second follows from observing that the highest-degree term in $H_n(x)$ is x^n and that $b_{n,n}=1$.
(2) By V-1.4.2,

$$\exp(\theta x)=\exp\left(\frac{\theta^2}{2}\right)\sum_{k=0}^{\infty}\theta^k\frac{H_k(x)}{k!}.$$

Hence

$$\begin{aligned}\int_{-\infty}^{+\infty}\exp(\theta t)\sigma(y,dt) &= \int_{-\infty}^{+\infty}\exp\left(\frac{\theta x}{y}\right)K(y,dx)\\ &= \exp\left(\tfrac{\theta^2}{2y^2}\right)\int_{-\infty}^{+\infty}\sum_{k=0}^{\infty}\frac{\theta^k}{k!}y^{-k}H_k(x)K(y,dx).\end{aligned}$$

That the integral and the summation sign can be interchanged is not completely clear. To simplify, we write

$$s_N(x)=\sum_{k=0}^{N}\frac{\theta^k}{k!}H_k(x)\quad\text{and}\quad\lim_{N\to\infty}s_N(x).$$

Then

$$\int\nu_1(dy)\int(s(x)-s_N(x))^2K(y,dx)=\int(s(x)-s_N(x))^2\nu_1(dx)=\sum_{i\geq N}\frac{\theta^{2i}}{i!}.$$

Thus, if $f_N(y)=\int(s(x)-s_N(x))^2K(y,dx)$, we see that $\lim_{N\to\infty}\mathbf{E}(f_N(Y))=0$. It then follows from Chebyshev's inequality (Problem IV-6) that $\nu_1(\{y:f_N(y)\geq\epsilon\})\to 0$ for all $\epsilon>0$. That is, $f_N(Y)\to 0$ in probability. Hence there exists a subsequence $\{N_n\}_{n\geq 0}$ such that $\lim_{n\to\infty}f_{N_n}(y)=0$ y-almost everywhere.

Since $\int |s(x) - s_{N_n}(x)| K(y, dx) \leq (f_{N_n}(y))^{\frac{1}{2}}$ by Schwarz's inequality,

$$\int \lim_{n\to\infty} \left(\sum_{n=0}^{N_n} \frac{\theta^k}{k!} H_k(x) \right) K(y, dx) = \int s(x) K(y, dx),$$

and finally

$$\int_{-\infty}^{+\infty} \exp(\theta t)\sigma(y, dt) = \exp \frac{\theta^2}{2y^2} \sum_{k=0}^{\infty} C_k \frac{\theta^k}{k!} y^{-k} H_k(y).$$

Since $y^{-k}H_k(y) \to 1$ as $y \to \infty$, a standard imitation of the method of Problem V-2 shows that $\sum_{k=N}^{\infty} C_k \frac{\theta^k}{k!} y^{-k} H_k(y)$ tends uniformly to 0 in y as $N \to +\infty$. Hence

$$\lim_{y\to\infty} \int \exp(\theta t)\sigma(y, dt) = \sum_{k=0}^{\infty} C_k \frac{\theta^k}{k!}$$

for every θ in $\mathbf{C}$.
(3) Applying Paul Lévy's theorem (with $\theta = it$, $t \in \mathbf{R}$), we see that the probability measure $\sigma(dt) = \lim_{y\to+\infty} \sigma(y, dt)$ exists and satisfies

$$\int e^{\theta x} \sigma(dt) = \sum_{k=0}^{\infty} C_k \frac{\theta^k}{k!}.$$

In particular, $\int t^k \sigma(dt) = C_k$.

To see that σ is concentrated on $[-1, +1]$, suppose that there exists $t_0 \notin [-1, 1]$ such that $p_\epsilon = \sigma((-x_0 - \epsilon, x_0 + \epsilon)) > 0$ for all $\epsilon > 0$. Then

$$(C_{2k})^{\frac{1}{2k}} = \left(\int t^{2k} \sigma(dt) \right)^{\frac{1}{2k}} \geq p_\epsilon^{\frac{1}{2k}} (|x_0| - \epsilon),$$

which contradicts the fact that $|C_n| \leq 1$ for all n.
(4) Since $\int_{-1}^{1} t^n (\sigma(dt) - \sigma'(dt)) = 0$ for every n, it is clear that $\int_{-1}^{1} f(t)(\sigma(dt) - \sigma'(dt)) = 0$ for every polynomial f. But the polynomials are dense in the continuous functions on $[-1, 1]$ in the uniform norm topology, and therefore $\int_{-1}^{1} f(t)(\sigma(dt) - \sigma'(dt)) = 0$ for every continuous function f.
(5) Let $\mu \mapsto \sigma(\mu)$ denote the mapping. By (4), if $\lambda \in [0, 1]$, then $\sigma(\lambda\mu_1 + (1 - \lambda)\mu_2) = \lambda\sigma(\mu_1) + (1 - \lambda)\sigma(\mu_2)$. The mapping is injective because σ determines $\{C_n\}_{n=0}^{\infty}$ and hence μ. (This last assertion follows since, for every n,

$$\int H_n(x) K(y, dx) = C_n H_n(y),$$

and this determines $K(y, dx)$ because

$$\sum_{n\geq 0} \frac{\theta^n}{n!} \int H_n(x) K(y, dx) = \exp \frac{\theta^2}{2} \int e^{\theta x} K(y, dx).)$$

To see that it is surjective, note that there exists μ_ρ such that $\sigma(\mu_\rho) = \delta_\rho$. If $\rho = 1$, then $\mu_1(dx, dy)$ is the distribution of (X, X), where the distribution of X is ν_1. Similarly, μ_{-1} is the distribution of $(X, -X)$. If $|\rho| < 1$, then $\mu(dx, dy)$ is the centered normal distribution on $\mathbf{R}^2$ with covariance matrix $\begin{pmatrix} 1 & \rho \\ \rho & 1 \end{pmatrix}$. Finally, the fact that $\mu_n \to \mu$ narrowly implies that $\sigma(\mu_n) \to \sigma(\mu)$, and this in turn implies that $\sigma(\mathcal{C})$ is closed in the topology of narrow convergence. Since $\sigma(\mathcal{C})$ contains the convex combinations of Dirac measures on $[-1, 1]$, it is the whole set of probability measures on $[-1, 1]$.

REMARK. This phenomenon was observed by O. Sarmanov (1966) and generalized by Tyan, Derin, and Thomas (1976).